Cabbages and Cauliflowers: How to Grow Them

By James John Howard Gregory

I0468513

CONTENTS.

CABBAGES AND CAULIFLOWERS.

OBJECT OF THIS TREATISE.

As a general, yet very thorough, response to inquiries from many of my customers about cabbage raising, I have aimed in this treatise to tell them all about the subject. The different inquiries made from time to time have given me a pretty clear idea of the many heads under which information is wanted; and it has been my aim to give this with the same thoroughness of detail as in my little work on Squashes. I have endeavored to talk in a very practical way, drawing from a large observation and experience, and receiving, in describing varieties, some valuable information from McIntosh's work, "The Book of the Garden."

THE ORIGIN OF CABBAGE.

Botanists tell us that all of the Cabbage family, which includes not only every variety of cabbage, Red, White, and Savoy, but all the cauliflower, broccoli, kale, and brussels sprouts, had their origin in the wild cabbage of Europe (Brassica oleracea), a plant with green, wavy leaves, much resembling charlock, found growing wild at Dover in England, and other parts of Europe. This plant, says McIntosh, is mostly confined to the sea-shore, and grows only on chalky or calcareous soils.

Thus through the wisdom of the Great Father of us all, who occasionally in his great garden allows vegetables to sport into a higher form of life, and grants to some of these sports sufficient strength of individuality to enable them to perpetuate themselves, and, at times, to blend their individuality with that of other sports, we have the heading cabbage in its numerous varieties, the creamy cauliflower, the feathery kale, the curled savoy. On my own grounds from a strain of seed that had been grown isolated for years, there recently came a plant that in its structure closely resembled Brussels Sprouts, growing about two feet in height, with a small head under each leaf. The cultivated cabbage was first introduced into England by the Romans, and from there nearly all the kinds cultivated in this country were originally brought. Those which we consider as peculiarly American varieties, have only been made so by years of careful improvement on the original imported sorts. The characteristics of these varieties will be given farther on.

WHAT A CABBAGE IS.

If we cut vertically through the middle of the head, we shall find it made up of successive layers of leaves, which grow smaller and smaller, almost ad infinitum. Now, if we take a fruit bud from an apple-tree and make a similar section of it, we shall find the same structure. If we observe the development of the two, as spring advances, we shall find another similarity (the looser the head the closer will be the resemblance),--the outer leaves of each will unwrap and unfold, and a flower stem will push out from each. Here we see that a cabbage is a bud, a seed bud (as all fruit buds may be termed, the production of seed being the primary object in nature, the fruit enclosing it playing but a secondary part), the office of the leaves being to cover, protect, and afterwards nourish the young seed shoot. The outer leaves which surround the head appear to have the same office as the leaves which surround the growing fruit bud, and that office closes with the first year, as does that of the leaves surrounding fruit buds, when each die and drop off. In my locality the public must have perceived more or less clearly the analogy between the heads of cabbage and the buds of trees, for when they speak of small heads they frequently call them "buds." That the close wrapped leaves which make the cabbage head and surround the seed germ, situated just in the middle of the head at the termination of the stump, are necessary for its protection and nutrition when young, is proved, I think, by the fact that those cabbages, the heads of which are much decayed, when set out for seed, no matter how sound the seed germ may be at the end of the stump, never make so large or healthy a seed shoot as those do the heads of which are sound; as a rule, after pushing a feeble growth, they die.

For this reason I believe that the office of the head is similar to and as necessary as that of the leaves which unwrap from around the blossom buds of our fruit trees. It is true that the parallel cannot be fully maintained, as the leaves which make up the cabbage head do not to an equal degree unfold (particularly is this true of hard heads); yet they exhibit a vitality of their own, which is seen in the deeper green color the outer leaves soon attain, and the change from tenderness to toughness in their structure: I think, therefore, that the degree of failure in the parallel may be measured by the difference between a higher and a lower form of organic life.

Some advocate the economy of cutting off a large portion of the heads when cabbages are set out for seed to use as food for stock. There is certainly a great temptation, standing amid acres of large, solid, heads in the early spring months, when green food of all kinds is scarce, to cut and use such an immense amount of rich food, which, to the inexperienced eye, appears to be utterly wasted if left to decay, dry, and fall to the ground; but, for the reason given above, I have never done so. It is possible that large heads may bear trimming to a degree without injury to the seed crop; yet I should consider this an experiment, and one to be tried with a good deal of caution.

SELECTING THE SOIL.

In some of the best cabbage-growing sections of the country, until within a comparatively few years it was the very general belief that cabbage would not do well on upland. Accordingly the cabbage patch would be found on the lowest tillage land of the farm. No doubt, the lowest soil being the richer from a gradual accumulation of the wash from the upland, when manure was but sparingly used, cabbage would thrive better there than elsewhere,--and not, as was generally held, because that vegetable needed more moisture than any other crop. Cabbage can be raised with success on any good corn land, provided such land is well manured; and there is no more loss in seasons of drouth on such land than there is in seasons of excessive moisture on the lower tillage land of the farm. I wish I could preach a very loud sermon to all my farmer friends on the great value of liberal manuring to carry crops successfully through the effects of a severe drouth. Crops on soil precisely alike, with but a wall to separate them, will, in a very dry season, present a striking difference,--the one being in fine vigor, and the other "suffering from drouth," as the owner will tell you; but, in reality, from want of food.

The smaller varieties of cabbage will thrive well on either light or strong soil, but the largest drumheads do best on strong soil. For the Brassica family, including cabbages, cauliflowers, turnips, etc., there is no soil so suitable as freshly turned sod, provided the surface is well fined by the harrow; it is well to have as stout a crop of clover or grass, growing on this sod, when turned under, as possible, and I incline to the belief that it would be a judicious investment to start a thick growth of these by the application of guano to the surface sufficiently long before turning the sod to get an extra growth of the clover or grass. If the soil be very sandy in character, I would advise that the variety planted be the Winnigstadt, which, in my experience, is unexcelled for making a hard head under almost any conditions, however unpropitious. Should the soil be naturally very wet it should be underdrained, or stump foot will be very likely to appear, which is death to all success.

PREPARING THE SOIL.

Should the soil be a heavy clay, a deep fall ploughing is best, that the frosts of winter may disintegrate it; and should the plan be to raise an early crop, this end will be promoted by fall ploughing, on any soil, as the land will thereby be made drier in early spring. In New England the soil for cabbages should be ploughed as deep as the subsoil, and the larger drumheads should be planted only on the deepest soil. If the season should prove a favorable one, a good crop of cabbage may be grown on sod broken up immediately after a crop of hay has been taken from it, provided plenty of fine manure is harrowed in. One great risk here is from the dry weather that usually prevails at that season, preventing the prompt germination of the seed, or rooting of the plants. It is prudent in such a case to have a good stock of plants, that such as die may be promptly replaced. It is wise to plant the seed for these a week earlier than the main crop, for when transplanted to fill the vacant places it will take about a week for them to get well rooted.

The manure may be spread on the surface of either sod or stubble land and ploughed under, or be spread on the surface after ploughing and thoroughly worked into the soil by the wheel harrow or cultivator. On ploughed sod I have found nothing so satisfactory as the class of wheel harrows, which not only cut the manure up fine and work it well under, but by the same operation cut and pulverize the turf until the sod may be left not over an inch in thickness. To do the work thus thoroughly requires a yoke of oxen or a pair of stout horses. All large stones and large pieces of turf that are torn up and brought to the surface should be carted off before making the hills.

THE MANURE.

Any manure but hog manure for cabbage,--barn manure, rotten kelp, night-soil, guano, fertilizers, wood ashes, fish, salt, glue waste, hen manure, slaughter-house manure. I have used all of these, and found them all good when rightly applied. If pure hog manure is used it is apt to produce that corpulent enlargement of the roots known in different localities as "stump foot," "underground head," "finger and thumb;" but I have found barn manure on which hogs have run, two hogs to each animal, excellent. The cabbage is the rankest of feeders, and to perfect the larger sort a most liberal allowance of the richest composts is required. To grow the smaller varieties either barn-yard manure, guano, fertilizers, or wood ashes, if the soil be in good condition, will answer; though the richer and more abundant the manure the larger are the cabbages, and the earlier the crop will mature.

To perfect the large varieties of drumhead,--by which I mean to make them grow to the greatest size possible,--I want a strong compost of barn-yard manure, with night-soil and muck or fish-waste, and, if possible, rotten kelp. A compost into which night-soil enters as a component is best made by first covering a plot of ground, of easy access, with soil or muck that has been exposed to a winter's frost, to the depth of about eighteen inches, and raising around this a rim about three feet in height, and thickness. Into this the night-soil is poured from carts built for the purpose, until the receptacle is about two-thirds full. Barn manure is now added, being dropped around and covering the outer rim, and, if the supply is sufficient, on the top of the heap also, on which it can be carted after cold weather sets in. Early in spring, the entire mass should be pitched over, thoroughly broken up with the bar and pick where frozen, and the frozen masses thrown on the surface. In pitching over the mass, work the rim in towards the middle of the heap. After the frozen lumps have thawed, give the heap another pitching over, aiming to mix all the materials thoroughly together, and make the entire mass as fine as possible. A covering of sand, thrown over the heap, before the last pitching, will help fine it.

To produce a good crop of cabbages, with a compost of this quality, from six to twelve cords will be required to the acre. If the land is in good heart, by previous high cultivation, or the soil is naturally very strong, six cords will give a fair crop of the small varieties; while, with the same conditions, from nine to twelve cords to the acre will be required to perfect the largest variety grown, the Marblehead Mammoth Drumhead.

Of the other kinds of manure named above, I will treat farther under the head of:

HOW TO APPLY THE MANURE.

The manure is sometimes applied wholly in the hill, at other times partly broadcast and partly in the hill. If the farmer desires to make the utmost use of his manure for that season, it will be best to put most of it into the hill, particularly if his supply runs rather short; but if he desires to leave his land in good condition for next year's crop, he had better use part of it broadcast. My own practice is to use all my rich compost broadcast, and depend on guano, fertilizers, or hen manure in the hill. Let all guano, if at all lumpy, like the Peruvian, be sifted, and let all the hard lumps be reduced by pounding, until the largest pieces shall not be larger than half a pea, before it is brought upon the ground. My land being ready, the compost worked under and the rows marked out, I select three trusty hands who can be relied upon to follow faithfully my directions in applying so dangerous manure as guano is in careless or ignorant hands; one takes a bucket of it, and, if for large cabbage, drops as much as he can readily close in his shut hand, where each hill is to be; if for small sorts, then about half that quantity, spreading it over a circle about a foot in diameter; the second man follows with a pronged hoe, or better yet, a six-tined fork, with which he works the guano well into the soil, first turning it three or four inches under the surface, and then stirring the soil very thoroughly with the hoe or fork. Unless the guano (and this is also true of most fertilizers) is faithfully mixed up with the soil, the seed will not vegetate. Give the second man about an hour the start, and then let the third man follow with the seed. Of other fertilizers, I use about half as much again as of guano to each hill, and of hen manure a heaping handful, after it has been finely broken up, and, if moist, slightly mixed with dry earth. When salt is used, it should not be depended on exclusively, but be used in connection with other manures, at the rate of from ten to fifteen bushels to the acre, applied broadcast over the ground, or thoroughly mixed with the manure before that is applied; if dissolved in the manure, better yet. Salt itself is not a manure. Its principal office is to change other materials into plant food. Fish and glue waste are exceedingly powerful manures, very rich in ammonia, and, if used the first season, they should be in compost. It is best to handle fish waste, such as heads, entrails, backbones, and liver waste, precisely like night soil. "Porgy cheese," or "chum," the refuse, after pressing out the oil from menhaden and halibut heads, and sometimes sold extensively for manure, is best prepared for use by composting it with muck or loam, layer with layer, at the rate of a barrel to every foot and a half, cord measure, of soil. As soon as it shows some heat, turn it, and repeat the process, two or three times, until it is well decomposed, when apply. Another excellent way to use fish waste is to compost it with barn manure, in the open fields. It will be best to have six inches of soil under the heap, and not layer the fish with the lower half of the manure, for it strikes down. Glue waste is a very coarse, lumpy manure, and requires a great deal of severe manipulation, if it is to be applied the first season. A better way is to compost it with soil, layer with layer, having each layer about a foot in thickness, and so allow it to remain over until the next season, before using. This will decompose most of the straw, and break down the hard, tough lumps. In applying this to the crop, most of it had better be used broadcast, as it is apt, at best, to be rather too coarse and concentrated to be used liberally directly in the hill. Slaughter-house manure should be treated much like glue manure.

Mr. Proctor, of Beverly, has raised cabbage successfully on strong clay soil, by spreading a compost of muck containing fish waste, in which the fish is well decomposed, at the rate of two tons of the fish to an acre of land, after plowing, and then, having made his furrows at the right distance apart, harrowing the land thoroughly crossways with the furrows. The result was, besides mixing the manure thoroughly with the soil, to land an extra proportion of it in the furrows, which was equivalent to manuring in the drill.

Cabbage can be raised on fertilizers alone. I have raised some crops in this way; but have been led to plow in from four to six cords of good manure to the acre, and then use from five hundred to a thousand pounds of some good fertilizer in the hill. The reason I prefer to use a portion of the cabbage food in the form of manure, is, that I have noticed that when the attempt is made to raise the larger drumhead varieties on fertilizers only, the cabbages, just as the heads are well formed, are apt to come nearly to a standstill. I explain this on the supposition that they exhaust most of the fertilizer, or some one of the ingredients that enter into it, during the earlier stage of growth; perhaps from the fact that the food is in so easily digestible condition, they use an over share of it, and the fact that those fed on fertilizers only, tend to grow longer stumped than usual, appears to give weight to this opinion. Though any good fertilizer is good for cabbage, yet I prefer those compounded on the basis of an analysis of the composition of the plants; they should contain the three ingredients, nitrogen, potash, and phosphoric acid, in the proportion of six, seven, five, taking them in the order in which I have written them.

MAKING THE HILLS AND PLANTING THE SEED.

The idea is quite prevalent that cabbages will not head up well except the plants are started in beds, and then transplanted into the hills where they are to mature. This is an error, so far as it applies to the Northern States,--the largest and most experienced cultivators of cabbage in New England usually dropping the seed directly where the plant is to stand, unless they are first started under glass, or the piece of land to be planted cannot be prepared in season to enable the farmer to put his seed directly in the hill and yet give the cabbage time sufficient to mature. Where the climate is unpropitious, or the quantity of manure applied is insufficient, it is possible that transplanting may promote heading. The advantages of planting directly in the hill, are a saving of time, avoiding the risks incidental to transplanting, and having all the piece start alike; for, when transplanted, many die and have to be replaced, while some hesitate much longer than others before starting, thus making a want of uniformity in the maturing of the crop. There is, also, this advantage, there being several plants in each hill, the cut-worm has to depredate pretty severely before he really injures the piece; again, should the seed not vegetate in any of the hills, every farmer will appreciate the advantage of having healthy plants growing so near at hand that they can be transferred to the vacant spaces with their roots so undisturbed that their growth is hardly checked. In addition to the labor of transplanting saved by this plan, the great check that plants always receive when so treated is prevented, and also the extra risks that occur should a season of drouth follow. It is the belief of some farmers, that plants growing where the seed was planted are less liable to be destroyed by the cut-worm than those that have been transplanted. When planning to raise late cabbage on upland, I sow a portion of the seed on a moist spot, or, in case a portion of the land is moist, I plant the hills on such land with an extra quantity of seed, that I may have enough plants for the whole piece, should the weather prove to be too dry for the seed to vegetate on the dryer portions of it. It is wise to sow these extra plants about a week earlier, for they will be put back about a week by transplanting them.

Some of our best farmers drill their seed in with a sowing machine, such as is used for onions, carrots, and other vegetable crops. This is a very expeditious way, and has the advantage of leaving the plants in rows instead of bunches, as in the hill system, and thus enables the hoe to do most of the work of thinning. It has also this advantage: each plant being by itself can be left much longer before thinning, and yet not grow long in the stump, thus making it available for transplanting, or for sale in the market, for a longer period.

The usual way of preparing the hills is to strike out furrows with a small, one-horse plough, as far apart as the rows are to be. As it is very important that the rows should be as straight as practicable, it is a good plan to run back once in each furrow, particularly on sod land where the plough will be apt to catch in the turf and jump out of line. A manure team follows, containing the dressing for the hills, which has previously been pitched over and beaten up until all the ingredients are fine and well mixed. This team is so driven, if possible, as to avoid running in the furrows. Two or three hands follow with forks or shovels, pitching the manure into the furrows at the distance apart that has been determined on for the hills. How far apart these are to be will depend on the varieties, from eighteen inches to four feet. On land that has been very highly manured for a series of years, cabbage can be planted nearer than on land that has been under the plow but a few years. For the distance apart for different varieties see farther on. The manure is levelled with hoes, a little soil is drawn over it, and a slight stamp with the back of the hoe is given to level this soil, and, at the same time, to mark the hill. The planter follows with seed in a tin box, or any small vessel having a broad bottom, and taking a small pinch between the thumb and forefinger he gives a slight scratch with the remaining fingers of the same hand, and dropping in about half a dozen seed covers them half an inch deep with a sweep of the hand, and packs the earth by a gentle pat with the open palm to keep the moisture in the ground and thus promote the vegetation of the seed. With care a quarter of a pound of seed will plant an acre, when dropped directly in the hills; but half a pound is the common allowance, as there is usually some waste from spilling, while most laborers plant with a free hand.

The soil over the hills being very light and porous, careless hands are apt to drop the seed too deep. Care should be taken not to drop the seed all in one spot, but to scatter them over a surface of two or three inches square, that each plant may have room to develop without crowding its neighbors.

If the seed is planted in a line instead of in a mass the plants can be left longer before the final thinning without danger of growing tall and weak.

If the seed is to be drilled in, it will be necessary to scatter the manure all along the furrows, then cover with a plough, roughly leveling with a rake.

Should the compost applied to the hills be very concentrated, it will be apt to produce stump foot; it will, therefore, be safest in such cases to hollow out the middle with the corner of the hoe, or draw the hoe through and fill in with earth, that the roots of the young plants may not come in direct contact with the compost as soon as they begin to push.

When guano or phosphates are used in the hills it will be well to mark out the rows with a plough, and then, where each hill is to be, fill in the soil level to the surface with a hoe, before applying them. I have, in a previous paragraph, given full instructions how to apply these. Hen manure, if moist, should be broken up very fine, and be mixed with some dry earth to prevent it from again lumping together, and the mixture applied in sufficient quantity to make an equivalent of a heaping handful of pure hen manure to each hill. Any liquid manure is excellent for the cabbage crop; but it should be well diluted, or it will be likely to produce stump foot.

Cabbage seed of almost all varieties are nearly round in form, but are not so spherical as turnip seed. I note, however, that seed of the Savoys are nearly oval. In color they are light brown when first gathered, but gradually turn dark brown if not gathered too early. An ounce contains nearly ten thousand seed, but should not be relied upon for many over two thousand good plants, and these are available for about as many hills only when raised in beds and transplanted; when dropped directly in the hills it will take not far from eight ounces of the larger sorts to plant an acre, and of the smaller cabbage rather more than this. Cabbage seed when well cured and kept in close bags will retain their vitality four or five years; old gardeners prefer seed of all the cabbage family two or three years old.

When the plan is to raise the young plants in beds to be transplanted, the ground selected for the beds should be of rich soil; this should be very thoroughly dug, and the surface worked and raked very fine, every stone and lump of earth being removed. Now sprinkle the seed evenly over the bed and gently rake in just under the surface, compacting the soil by pressure with a board. As soon as the young plants appear, sprinkle them with air-slaked lime. Transplant when three or four inches high, being very careful not to let the plants get tall and weak.

For late cabbage, in the latitude of Boston, to have cabbages ready for market about the first of November, the Marblehead Mammoth should be planted the 20th of May, other late drumheads from June 1st to June 12th, provided the plants are not to be transplanted; otherwise a week earlier. In those localities where the growing season is later, the seed should be planted proportionally later.

CARE OF THE YOUNG PLANTS.

In four or five days, if the weather is propitious, the young plants will begin to break ground, presenting at the surface two leaves, which together make nearly a square, like the first leaves of turnips or radishes. As soon as the third leaf is developed, go over the piece, and boldly thin out the plants. Wherever they are very thick, pull a mass of them with the fingers and thumb, being careful to fill up the hole made with fine earth. After the fourth leaf is developed, go over the piece again and thin still more; you need specially to guard against a slender, weak growth, which will happen when the plants are too crowded. In thinning, leave the short-stumped plants, and leave them as far apart in the hill as possible, that they may not shade each other, or so interfere in growing as to make long stumps. If there is any market for young plants, thousands can be sold from an acre when the seed are planted in the hill; but in doing this bear in mind that your principal object is to raise cabbages, and to succeed in this the young plants must on no account be allowed to stand so long together in the hills as to crowd each other, making a tall, weak, slender growth,--getting "long-legged," as the farmers call it.

If the manure in any of the hills is too strong, the fact will be known by its effects on the plants, which will be checked in their growth, and be of a darker green color than the healthy plants. Gently pull away the earth from the roots of such with the fingers, and draw around fresh earth; or, what is as well or better, transplant a healthy plant just on the edge of the hill. When the plants are finger high they are of a good size to transplant into such hills as have missed, or to market. When transplanting, select a rainy day, if possible, and do not begin until sufficient rain has fallen to moisten the earth around the roots, which will make it more likely to adhere to them when taken up. Take up the young plants by running the finger or a trowel under them; put these into a flat basket or box, and in transplanting set them to the same depth they originally grew, pressing the earth a little about the roots.

If it is necessary to do the transplanting in a dry spell, as usually happens, select the latter part of the afternoon, if practicable, and, making holes with a dibble, or any pointed stick an inch and a half in diameter, fill these holes, a score or more at a time, with water; and as soon as the water is about soaked away, beginning with the hole first filled, set out your plants. The evaporation of the moisture below the roots will keep them moist until they get a hold. Cabbage plants have great tenacity of life, and will rally and grow when they appear to be dead; the leaves may all die, and dry up like hay, but if the stump stands erect and the unfolded leaf at the top of the stump is alive, the plant will usually survive. When the plants are quite large, they may be used successfully by cutting or breaking off the larger leaves. Some advocate wilting the plants before transplanting, piling them in the cellar a few days before setting them out, to toughen them and get a new setting of fine roots; others challenge their vigor by making it a rule to do all transplanting under the heat of mid-day. I think there is not much of reason in this latter course. The young plants can be set out almost as fast as a man can walk, by holding the roots close to one side of the hole made by the dibble, and at the same moment pressing earth against them with the other hand.

PROTECTING THE PLANTS FROM THEIR ENEMIES.

As soon as they have broken through the soil, an enemy awaits them in the small black insect commonly known as the cabbage or turnip fly, beetle, or flea. This insect, though so small as to appear to the eye as a black dot, is very voracious and surprisingly active. He apparently feeds on the juice of the young plant, perforating it with small holes the size of a pin point. He is so active when disturbed that his motions cannot be followed by the eye, and his sense of danger is so keen that only by cautiously approaching the plant can he be seen at all. The delay of a single day in protecting the young plants from his ravages will sometimes be the destruction of nearly the entire piece. Wood ashes and air-slaked lime, sprinkled upon the plants while the leaves are moist from either rain or dew, afford almost complete protection. The lime or ashes should be applied as soon as the plant can be seen, for then, when they are in their tenderest condition, the fly is most destructive. I am not certain that the alkaline nature of these affords the protection, or whether a mere covering by common dust might not answer equally well. Should the covering be washed off by rain, apply it anew immediately after the rain has ceased, and so continue to keep the young plants covered until the third or fourth leaves are developed when they will have become too tough to serve as food for this insect enemy.

A new enemy much dreaded by all cabbage raisers will begin to make his appearance about the time the flea disappears, known as the cut-worm. This worm is of a dusky brown color, with a dark colored head, and varies in size up to about two inches in length. He burrows in the ground just below the surface, is slow of motion, and does his mischievous work at night, gnawing off the young plants close at the surface of the ground. This enemy is hard to battle with. If the patch be small, these worms can be scratched out of their hiding places by pulling the earth carefully away the following morning for a few inches around the stump of the plant destroyed, when the rascals will usually be found half coiled together. Dropping a little wood ashes around the plants close to the stumps is one of the best of remedies; its alkaline properties burning his nose I presume. A tunnel of paper put around the stump but not touching it, and sunk just below the surface, is recommended as efficacious; and from the habits of the worm I should think it would prove so. Perpendicular holes four inches deep and an inch in diameter is said to catch and hold them as effectively as do the pit falls of Africa the wild animals. Late planted cabbage will suffer little or none from this pest, as he disappears about the middle of June. Some seasons they are remarkably numerous, making it necessary to replant portions of the cabbage patch several times over. I have heard of as many as twenty being dug at different times the same season out of one cabbage hill. The farmer who tilled that patch earned his dollars. When the cabbage has a stump the size of a pipe stem it is beyond the destructive ravages of the cut-worm, and should it escape stump foot has usually quite a period of growth free from the attacks of enemies. Should the season prove unpropitious and the plant be checked in its growth, it will be apt to become "lousy," as the farmers term it, referring to its condition when attacked by a small green insect known as aphid? which preys upon it in myriads; when this is the case the leaves lose their bright green, turn of a bluish cast, the leaf stocks lose somewhat of their supporting powers, the leaves curl up into irregular shapes, and the lower layer turns black and drops off, while the ground under the plant appears covered with the casts or bodies of the insects as with a white powder. When in this condition the plants are in a very bad way.

Considering the circumstances under which this insect appears, usually in a very dry season, I hold that it is rather the product than the cause of disease, as with the bark louse on our apple-trees; as a remedy I advocate sprinkling the plants with air-slaked lime, watering, if possible, and a frequent and thorough stirring of the soil with the cultivator and hoe. The better the opportunities the cabbage have to develop themselves through high manuring, sufficient moisture, good drainage, and thorough cultivation, the less liable they are to be "lousy." As the season advances there will sometimes be found patches eaten out of the leaves, leaving nothing but the skeleton of leaf veins; an examination will show a band of caterpillars of a light green color at work, who feed in a compact mass, oftentimes a square, with as much regularity as though under the best of military discipline. The readiest way to dispose of them is to break off the leaf and crush them under foot. The common large red caterpillar occasionally preys on the plants, eating large holes in the leaves, especially about the head. When the cabbage plot is bordered by grass land, in seasons when grasshoppers are plenty, they will frequently destroy the outer rows, puncturing the leaves with small holes, and feeding on them until little besides their skeletons remain. In isolated locations rabbits and other vegetable feeders sometimes commit depredations. The snare and the shot-gun are the remedy for these.

Other insects that prey upon the cabbage tribe, in their caterpillar state, are the cabbage moth, white-line, brown-eyed moth, large white garden butterfly, white and green veined butterfly. All of these produce caterpillars, which can be destroyed either by application of air-slaked lime, or by removing the leaves infested and crushing the intruders under foot. The cabbage-fly, father-long-legs, the millipedes, the blue cabbage-fly, brassy cabbage-flea, and two or three other insect enemies are mentioned by McIntosh as infesting the cabbage fields of England; also three species of fungi known as white rust, mildew, and cylindrosporium concentricum; these last are destroyed by the sprinkling of air-slaked lime on the leaves. In this country, along the sea coast of the northern section, in open-ground cultivation, there is comparatively but little injury done by these marauders, which are the cause of so much annoyance and loss to our English cousins.

THE GREEN WORM.

A new and troublesome enemy to the cabbage tribe which has made its appearance within a few years, and spread rapidly over a large section of the country, is a green worm, Anthomia brassic? This pest infests the cabbage tribe at all stages of its growth; it is believed to have been introduced into this country from Europe, by the way of Canada, where it was probably brought in a lot of cabbage. It is the caterpillar of a white butterfly with black spots on its wings. In Europe, this butterfly is preyed on by two or more parasites, which keep it somewhat in check; but its remarkably rapid increase in this country, causing a wail of lamentation to rise in a single season from the cabbage growers over areas of tens of thousands of square miles, proved that when it first appeared it had reached this country without its attendant parasites.

Besides this green worm, there are found in Europe four varieties of caterpillar variously marked, the caterpillars from all of which make great havoc among the cabbage tribe.

The most effective destroyer of this, and about every other insect pest, is what is known as the "Kerosene Emulsion." This is made by churning common kerosene with milk or soap until it is diffused through the liquid.

Take one quart of kerosene oil and pour it into a pint of hot water in which an ounce of common soap has been dissolved; churn this briskly while hot (a force pump is excellent for this), and, when well mixed, which will be in a few minutes, it will be of a creamy consistency; mix one quart to ten or twelve of cold water, and spray or sprinkle it over the plants with a force-pump syringe or a whisk broom.

Another remedy is pyrethrum. Use that which is fresh; either blowing it on in a dry state with a bellows, wherever the worm appears, or using it diluted, at the rate of a tablespoonful to two gallons of water; applying as with the kerosene emulsion. Mr. A. S. Fuller, who is good authority on garden matters, succeeds by applying tar-water. Place a couple of quarts of coal tar in a barrel and fill with water; let it stand forty-eight hours, then dip off, and apply with a watering-pot, or syringe.

Chickens allowed to run freely among the growing plants, the hen being confined in a movable coop, if once attracted to them will fatten on them. This remedy might answer very well for small plots. Large areas in cabbage, in proportion to their size are, as a rule, far less injured by insect enemies than small patches. The worm is of late years less troublesome in the North than formerly.

CLUB OR STUMP FOOT AND MAGGOT.

The great dread of every cabbage grower is a disease of the branching roots, producing a bunchy, gland-like enlargement, known in different localities under the name of club foot, stump foot, underground head, finger and thumb. The result is a check in the ascent of the sap, which causes a defective vitality. There are two theories as to the origin of club foot; one that it is a disease caused by poor soil, bad cultivation, and unsuitable manures; the other that the injury is done by an insect enemy, Curculio contractus. It is held by some that the maggots at the root are the progeny of the cabbage flea. This I doubt. This insect, "piercing the skin of the root, deposits its eggs in the holes, lives during a time on the sap of the plant, and then escapes and buries itself for a time in the soil."

If the wart, or gland-like excrescence, is seen while transplanting, throw all such plants away, unless your supply is short; in such case, carefully trim off all the diseased portions with a sharp knife. If the disease is in the growing crop, it will be made evident by the drooping of the leaves under the mid-day sun, leaves of diseased plants drooping more than those of healthy ones, while they will usually have a bluer cast. Should this disease show itself, set the cultivator going immediately, and follow with the hoe, drawing up fresh earth around the plants, which will encourage them to form new fibrous roots; should they do this freely, the plants will be saved, as the attacks of the insect are usually confined to the coarse, branching roots. Should the disease prevail as late as when the plants have reached half their growth, the chances are decidedly against raising a paying crop.

When the land planted is too wet, or the manure in the hill is too strong, this dreaded disease is liable to be found on any soil; but it is most likely to manifest itself on soils that have been previously cropped with cabbage, turnip, or some other member of the Brassica family.

Farmers find that, as a rule, it is not safe to follow cabbage, ruta baga, or any of the Brassica family, with cabbage, unless three or four years have intervened between the crops; and I have known an instance in growing the

Marblehead Mammoth, where, though five years had intervened, that portion of the piece occupied by the previous crop could be distinctly marked off by the presence of club-foot.

Singular as it may appear, old gardens are an exception to this rule. While it is next to impossible to raise, in old gardens, a fair turnip, free from club-foot, cabbages may be raised year after year on the same soil with impunity, or, at least, with but trifling injury from that disease. This seems to prove, contrary to English authority, that club-foot in the turnip tribe is the effect of a different cause from the same disease in the cabbage family.

There is another position taken by Stephens in his "Book of the Farm," which facts seem to disprove. He puts forth the theory that "all such diseases arise from poverty of the soil, either from want of manure when the soil is naturally poor, or rendered effete by over-cropping." There is a farm on a neck of land belonging to this town (Marblehead, Mass.), which has peculiar advantages for collecting sea kelp and sea moss, and these manures are there used most liberally, particularly in the cultivation of cabbage, from eight to twelve cords of rotten kelp, which is stronger than barn manure, and more suitable food for cabbage, being used to the acre. A few years ago, on a change of tenants, the new incumbent heavily manured a piece for cabbage, and planted it; but, as the season advanced, stump-foot developed in every cabbage on one side of the piece, while all the remainder were healthy. Upon inquiry, he learned that, by mistake, he had overlapped the cabbage plot of last season just so far as the stump-foot extended. In this instance, it could not have been that the cabbage suffered for want of food; for, not only was the piece heavily manured that year and the year previous, but it had been liberally manured through a series of years, and, to a large extent, with the manure which, of all others, the cabbage tribe delight in, rotten kelp and sea mosses. I have known other instances where soil, naturally quite strong, and kept heavily manured for a series of years, has shown stump-foot when cabbage were planted, with intervals of two and three years between. My theory is, that the mere presence of the cabbage causes stump-foot on succeeding crops grown on the same soil. This is proved by the fact that where a piece of land in grass, close adjoining a piece of growing cabbage, had been used for stripping them for market, when this was broken up the next season and planted to cabbage, stump-foot appeared only on that portion where the waste leaves fell the year previous. I have another instance to the same point, told me by an observing farmer, that, on a piece of sod land, on which he ran his cultivator the year previous, when turning his horse every time he had cultivated a row, he had stump-footed cabbage the next season just as far as that cultivator went, dragging, of course, a few leaves and a little earth from the cabbage piece with it. Still, though the mere presence of cabbage causes stump-foot, it is a fact, that, under certain conditions, cabbage can be grown on the same piece of land year after year successfully, with but very little trouble from stump-foot. In this town (Marblehead), though, as I have stated, we cannot, on our farms, follow cabbage with cabbage, even with the highest of manuring and cultivation, yet in the gardens of the town, on the same kind of soil (and our soil is green stone and syenite, not naturally containing lime), there are instances where cabbage has been successfully followed by cabbage, on the same spot, for a quarter of a century and more. In the garden of an aged citizen of this town, cabbages have been raised on the same spot of land for over half a century.

The cause of stump foot cannot, therefore, be found in the poverty of the soil, either from want of manure or its having been rendered effete from over cropping. It is evident that by long cultivation soils gradually have diffused through them something that proves inimical to the disease that produces stump foot. I will suggest as probable that the protection is afforded by the presence of some alkali that old gardens are constantly acquiring through house waste which is always finding its way there, particularly the slops from the sink, which abound in potash. This is rendered further probable from the fact given by Mr. Peter Henderson, that, on soils in this vicinity, naturally abounding in lime, cabbage can be raised year following year with almost immunity from stump foot. He ascribes this to the effects of lime in the soil derived from marine shells, and recommends that lime from bones be used to secure the same protection; but the lime that enters into the composition of marine shells is for the most part carbonate of lime, whereas the greater portion of that which enters into the composition of bones is phosphate of lime. Common air-slaked lime is almost pure carbonate of lime, and hence comes nearer to the composition of marine shells than lime from bones, and, being much cheaper, would appear to be preferable.

An able farmer told me that by using wood ashes liberally he could follow with cabbage the next season on the same piece. One experiment of my own in this direction did not prove successful, where ashes at the rate of two hundred bushels to the acre were used; and I have an impression that I have read of a like want of success after quite liberal applications of lime. In a more recent experiment, on a gravelly loam on one of my seed farms in Middleton, Mass., where two hundred bushels of unleached ashes were used per acre, three-fourths broadcast, I have had complete success, raising as good a crop as I ever grew the second year on the same land, without a single stump foot on half an acre. Still, it remains evident, I think, that nature prevents stump foot by the diffusing of alkalies through the soil, and I mistrust that the reason why we sometimes fail with the same remedies is that we have them mixed, rather than intimately combined, with the particles of soil.

The roots of young plants are sometimes attacked by a maggot, though there is no club root present. A remedy for this is said to be in the burying of a small piece of bi-sulphide of carbon within a few inches of the diseased plant. I have never tried it, but know that there is no better insecticide.

As I have stated under another head, an attack of club foot is almost sure to follow the use of pure hog manure, whether it be used broadcast or in the hill. About ten years ago I ventured to use hog manure nearly pure, spread broadcast and ploughed in. Stump foot soon showed itself. I cultivated and hoed the cabbage thoroughly; then, as they still appeared sickly, I had the entire piece thoroughly dug over with a six-tined fork, pushing it as deep or deeper into the soil than the plough had gone, to bring up the manure to the surface; but all was of no use; I lost the entire crop. Yet, on another occasion, stable manure on which hogs had been kept at the rate of two hogs to each animal, gave me one of the finest lots of cabbage I ever raised.

CARE OF THE GROWING CROP.

As soon as the young plants are large enough to be seen with the naked eye, in with the cultivator and go and return once in each row, being careful not to have any lumps of earth cover the plants. Follow the cultivator immediately with the hoe, loosening the soil about the hills. The old rule with farmers is to cultivate and hoe cabbage three times during their growth, and it is a rule that works very well where the crop is in good growing condition; but if the manure is deficient, the soil bakes, or the plants show signs of disease, then cultivate and hoe once or twice extra. "Hoe cabbage when wet," is another farmer's axiom. In a small garden patch the soil may be stirred among the plants as often as may be convenient: it can do no harm; cabbages relish tending, though it is not necessary to do this every day, as one enthusiastic cultivator evidently thought, who declared that, by hoeing his cabbages every morning, he had succeeded in raising capital heads.

If a season of drouth occurs when the cabbages have begun to head, the heads will harden prematurely; and then should a heavy rain fall, they will start to make a new growth, and the consequence will be many of them will split. Split or bursted cabbage are a source of great loss to the farmer, and this should be carefully guarded against by going frequently over the piece when the heads are setting, and starting every cabbage that appears to be about mature. A stout-pronged potato hoe applied just under the leaves, and a pull given sufficient to start the roots on one side, will accomplish what is needed. If cabbage that have once been started seem still inclined to burst, start the roots on the other side. Instead of a hoe they may be pushed over with the foot, or with the hand. Frequently, heads that are thus started will grow to double the size they had attained when about to burst. There is a marked difference in this habit in different varieties of cabbage. I find that the Hard-heading is less inclined to burst its head than any of the kinds I raise.

MARKETING THE CROP.

When preparing for market cabbages that have been kept over winter, particularly if they are marketed late in the season, the edges of the leaves of some of the heads will be found to be more or less decayed; do not strip such leaves off, but with a sharp knife cut clean off the decayed edges. The earlier the variety the sooner it needs to be marketed, for, as a rule, cabbages push their shoots in the spring in the order of their earliness. If they have not been sufficiently protected from the cold, the stumps will often rot off close to the head, and sometimes the rot will include the part of the stump that enters the head. If the watery-looking portion can be cut clean out, the head is salable; otherwise it will be apt to have an unpleasant flavor when cooked. As a rule, cabbages for marketing should be trimmed into as compact a form as possible; the heads should be cut off close to the stump, leaving two or three spare leaves to protect them. They may be brought out of the piece in bushel baskets, and be piled on the wagon as high as a hay stack, being kept in place by a stout canvas sheet tied closely down. In the markets of Boston, in the fall of the year, they are usually sold at a price agreed upon by the hundred head; this will vary not only with the size and quality of the cabbage, but with the season, the crop, and the quality in market on that particular day. Within a few years I have known the range of price for the Stone Mason or Fottler cabbage, equal in size and quality, to be from $3 to $17 per hundred; for the Marblehead Mammoth from $6 to $25 per hundred. Cabbages brought to market in the spring are usually sold by weight or by the barrel, at from $1 to $4 per hundred pounds.

The earliest cabbages carried to market sometimes bring extraordinary prices; and this has created a keen competition among market gardeners, each striving to produce the earliest, a difference of a week in marketing oftentimes making a difference of one half in the profits of the crop. Capt. Wyman, who controlled the Early Wyman cabbage for several years, sold some seasons thirty thousand heads if my memory serves me, at pretty much his own price. As a rule, it is the very early and the very late cabbages that sell most profitably. Should the market for very late cabbages prove a poor one, the farmer is not compelled to sell them, no matter at what sacrifice, as would be the case a month earlier; he can pit them, and so keep them over to the early spring market which is almost always a profitable one. In marketing in spring it should be the aim to make sale before the crops of spring greens become plenty, as these replace the cabbage on many tables. By starting cabbage in hot beds a crop of celery or squashes may follow them the same season.

KEEPING CABBAGES THROUGH THE WINTER.

In the comparatively mild climate of England, where there are but few days in the winter months that the ground remains frozen to any depth, the hardy cabbage grows all seasons of the year, and turnips left during winter standing in the ground are fed to sheep by yarding them over the different portions of the field. With the same impunity, in the southern portion of our own country, the cabbages are left unprotected during the winter months; and, in the warmer portions of the South they are principally a winter crop. As we advance farther North, we find that the degree of protection needed is afforded by running the plough along each side of the rows, turning the earth against them, and dropping a little litter on top of the heads. As we advance still farther northward, we find sufficient protection given by but little more than a rough roof of boards thrown over the heads, after removing the cabbages to a sheltered spot and setting them in the ground as near together as they will stand without being in contact, with the tops of the heads just level with the surface.

In the latitude of central New England, cabbages are not secure from injury from frost with less than a foot of earth thrown over the heads. In mild winters a covering of half that depth will be sufficient; but as we have no prophets to foretell our mild winters, a foot of earth is safer than six inches. Where eel-grass can be procured along the sea coast, or there is straw or coarse hay to spare, the better plan is to cover with about six inches of earth, and when this is frozen sufficiently hard to bear a man's weight (which is usually about Thanksgiving time), to scatter over it the eel-grass, forest leaves, straw, or coarse hay, to the depth of another six inches. Eel-grass, which grows on the sandy flats under the ocean along the coast, is preferred to any other covering as it lays light and keeps in dead air which is a non-conductor of heat. Forest leaves are next in value; but snow and water are apt to get among these and freezing solid destroy most of their protecting value. When I use forest leaves, I cover them with coarse hay, and add branches of trees to prevent its being blown away. In keeping cabbages through the winter, three general facts should be borne in mind, viz.: that repeated freezing and thawing will cause them to rot; that excessive moisture or warmth will also cause rot; while a dry air, such as is found in most cellars, will abstract moisture from the leaves, injure the flavor of the cabbage, and cause some of the heads to wilt, and the harder heads to waste. In the Middle States we have mostly to fear the wet of winter, and the plan for keeping for that section should, therefore, have particularly in view protection from moisture, while in the Northern States we have to fear the cold of winter, and, consequently, our plan must there have specially in view protection from cold.

When storing for winter, select a dry day, if possible, sufficiently long after rainy weather to have the leaves free of water,--otherwise they will spout it on to you, and make you the wettest and muddiest scarecrow ever seen off a farm,--then strip all the outer leaves from the head but the two last rows, which are needed to protect it. This may be readily done by drawing in these two rows toward the head with the left hand, while a blow is struck against the remaining leaves with the fist of the right hand. Next pull up the cabbages, which, if they are of the largest varieties, may be expeditiously done by a potato hoe. If they are not intended for seed purposes, stand the heads down and stumps up until the earth on the roots is somewhat dry, when it can be mostly removed by sharp blows against the stump given with a stout stick. In loading do not bruise the heads. Select the place for keeping them in a dry, level location, and, if in the North, a southern exposure, where no water can stand and there can be no wash. To make the pit, run the plough along from two to four furrows, and throw out the soil with the shovel to the requisite depth, which may be from six to ten inches; now, if the design is to roof over the pit, the cabbages may be put in as thickly as they will stand; if the heads are solid they may be either head up or stump up, and two layers deep; but if the heads are soft, then heads up and one deep, and not crowded very close, that they may have room to make heads during the winter. Having excavated an area twelve by six feet, set a couple of posts in the ground midway at each end, projecting about five feet above the surface; connect the two by a joist secured firmly to the top of each, and against this, extending to the ground just outside the pit, lay slabs, boards or poles, and cover the roof that will be thus formed with six inches of straw or old hay, and, if in the North, throw six or eight inches of earth over this. Leave one end open for entrance and to air the pit, closing the other end with straw or hay. In the North close both ends, opening one of them occasionally in mild weather.

When cabbages are pitted on a large scale this system of roofing is too costly and too cumbersome. A few thousand may be kept in a cool root cellar, by putting one layer heads down, and standing another layer heads up between these. Within a few years farmers in the vicinity of Lowell, Mass., have preserved their cabbages over winter, on a large scale, by a new method, with results that have been very satisfactory. They cut off that portion of the stump which contains the root; strip off most of the outer leaves, and then pile the cabbages in piles, six or eight feet high, in double rows, with boards to keep them apart, in cool cellars, which are built half out of ground. The temperature of these, by the judicious opening and closing of windows, is kept as nearly as possibly at the freezing point. The common practice in the North, when many thousands are to be stored for winter and spring sales, is to select a southern exposure having the protection of a fence or wall, if practicable, and, turning furrows with the plough, throw out the earth with shovels, to the depth of about six inches; the cabbages, stripped as before described, are then stored closely together, and straw or coarse hay is thrown over them to the depth of a foot or eighteen inches. Protected thus they are accessible for market at any time during the winter. If the design is to keep them over till spring, the covering may be first six inches of earth, to be followed, as cold increases, with six inches of straw, litter, or eel-grass. This latter is my own practice, with the addition of leaving a ridge of earth between every three or four rows, to act as a support and keep the cabbages from falling over. I am, also, careful to bring the cabbages to the pit as soon as pulled, with the earth among the roots as little disturbed as possible; and, should the roots appear to be dry, to throw a little earth over them after the cabbages are set in the trench. The few loose leaves remaining will prevent the earth from sifting down between the heads, and the air chambers thus made answer a capital purpose in keeping out the cold, as air is one of the best non-conductors of heat. It is said that muck-soil, when well drained, is an excellent one to bury cabbage in, as its antiseptic properties preserve them from decay. If the object is to preserve the cabbage for market purposes only, the heads may be buried in the same position in which they grew, or they may be inverted, the stump having no value in itself; but if for seed purposes, they must be buried head up, as, whatever injures the stump, spoils the whole cabbage for that object. I store between ten and fifty thousand heads annually to raise seed from, and carry them through till planting time with a degree of success varying from a loss, for seed purposes, of from one-half to thirty-three per cent. of the number buried; but, if handled early in spring, many that would be worthless for seed purposes, could be profitably marketed. A few years since, I buried a lot with a depth varying from one to four feet, and found, on uncovering them in the spring, that all had kept, and apparently equally well. In the winter of 1868, excessively cold weather came very early and unexpectedly, before my cabbage plot had received its full covering of litter. The consequence was, the frost penetrated so deep that it froze through the heads into the stumps, and, when spring came, a large portion of them came out spoiled for seed purposes, though most of them sold readily in the market. A cabbage is rendered worthless for seed when the frost strikes through the stump where it joins the head; and though, to the unpractised eye, all may appear right, yet, if the heart of the stump has a water-soaked appearance on being cut into, it will almost uniformly decay just below the head in the course of a few weeks after having been planted out. If there is a probability that the stumps have been frozen through, examine the plot early, and, if it proves so, sell the cabbages for eating purposes, no matter how sound and handsome the heads look; if you delay until time for planting out the cabbage for seed, meanwhile much waste will occur. I once lost heavily in Marblehead Mammoth cabbage by having them buried on a hill-side with a gentle slope. In the course of the winter they fell over on their sides, which let down the soil from above, and, closing the air-chambers between them, brought the huge heads into a mass, and the result was, a large proportion of them rotted badly. At another time, I lost a whole plot by burying them in soil between ledges of rock, which kept the ground very wet when spring opened; the consequence was, every cabbage rotted. If the heads are frozen more than two or three leaves deep before they are pitted they will not come out so handsome in the spring; but cabbages are very hardy, and they readily rally from a little freezing, either in the open ground or after they are buried, though it is best, when they are frozen in the open ground, to let them remain there until the frost comes out before removing them, if it can be done without too much risk of freezing still deeper, as they handle better then, for, being tougher, the leaves are not so easily broken. If the soil is frozen to any depth before the cabbages are removed, the roots will be likely to be injured in the pulling, a matter of no consequence if the cabbages are intended for market, but of some importance if they are for seed raising. Large cabbages are more easily pulled by giving them a little twist; if for seed purposes, this should be avoided, as it injures the stump. A small lot, that are to be used within a month, can be kept hung up by the stump in the cellar of a dwelling-house; they will keep in this way until spring; but the outer leaves will dry and turn yellow, the heads shrink some in size, and be apt to lose in quality. Some practise putting clean chopped straw in the bottom of a box or barrel, wetting it, and covering with heads trimmed ready for cooking, adding again wet straw and a layer of heads, so alternating until the barrel or box is filled, after which it is headed up and kept in a cool place, at, or a little below, the freezing point. No doubt this is an excellent way to preserve a small lot, as it has the two essentials to success, keeping them cool and moist.

Instead of burying them in an upright position, after a deep furrow has been made the cabbages are sometimes laid on their sides two deep, with their roots at the bottom of the furrow, and covered with earth in this position. Where the winter climate is so mild that a shallow covering will be sufficient protection, this method saves much labor.

HAVING CABBAGE MAKE HEADS IN WINTER.

When a piece of drumhead has been planted very late (sometimes they are planted on ground broken up after a crop of hay has been taken from it the same season), there will be a per cent. of the plants when the growing season is over that have not headed. With care almost all of these can be made to head during the winter. A few years ago I selected my seed heads from a large piece and then sold the first "pick" of what remained at ten cents a head, the second at eight cents, and so down until all were taken for which purchasers were willing to give one cent each. Of course, after such a thorough selling out as this, there was not much in the shape of a head left. I now had what remained pulled up and carted away, doubtful whether to feed them to the cows or to set them out to head up during winter. As they were very healthy plants in the full vigor of growth, having rudimentary heads just gathering in, I determined to set them out. I had a pit dug deep enough to bring the tops of the heads, when the plants were stood upright as they grew, just above the surface of the ground; I then stood the cabbages in without breaking off any of the leaves, keeping the roots well covered with earth, having the plants far enough apart not to crowd each other very much, though so near as to press somewhat together the two outer circles of leaves. They were allowed to remain in this condition until it was cold enough to freeze the ground an inch in thickness, when a covering of coarse hay was thrown over them a couple of inches thick, and, as the cold increased in intensity, this covering was increased to ten or twelve inches in thickness, the additions being made at two or three intervals. In the spring I uncovered the lot, and found that nearly every plant had headed up. I sold the heads for four cents a pound; and these refuse cabbages averaged me about ten cents a head, which was the price my best heads brought me in the fall. I have seen thousands of cabbages in one lot, the refuse of several acres that had been planted on sod land broken up the same season a crop of hay had been taken from it, made to head by this course, and sold in the spring for $1.30 per barrel. When there is a large lot of such cabbages the most economical way to plant them will be in furrows made by the plough. Most of the bedding used in covering them, if it be as coarse as it ought to be to admit as much air as possible while it should not mat down on the cabbages, will, with care in drying, be again available for covering another season, or remain suitable for bedding purposes. These "winter-headed" cabbages, as they are called in the market, are not so solid and have more shrinkage to them than those headed in the open ground; hence they will not bear transportation as well, neither will they keep as long when exposed to the air. The effect of wintering cabbage by burying in the soil is to make them exceedingly tender for table use.

VARIETIES OF CABBAGE.

If a piece of land is planted with seed grown from two heads of cabbage the product will bear a striking resemblance to the two parent cabbages, with a third variety which will combine the characteristics of these two, yet the resemblance will be somewhat modified at times by a little more manure, a little higher culture, a little better location, and the addition of an individuality that particular vegetables occasionally take upon themselves which we designate by the word "sport." The "sports" when they occur are fixed and perpetuated with remarkable readiness in the cabbage family, as is proved by a great number of varieties in cultivation, which are the numerous progeny of one ancestor. The catalogues of the English and French seedsmen contain long lists of varieties, many of which (and this is especially true of the early kinds) are either the same variety under a different name or are different "strains" of the same variety produced by the careful selections of prominent market gardeners through a series of years.

Every season I experiment with foreign and American varieties of cabbage to learn the characteristics of the different kinds, their comparative earliness, size, shape, and hardness of head, length of stump, and such other facts as would prove of value to market gardeners. There is one fact that every careful experimenter soon learns, that one season will not teach all that can be known relative to a variety, and that a number of specimens of each kind must be raised to enable one to make a fair comparison. It is amusing to read the dicta which appear in the agricultural press from those who have made but a single experiment with some vegetable; they proclaim more after a single trial than a cautious experimenter would dare to declare after years spent in careful observation. The year 1869 I raised over sixty varieties of cabbage, importing nearly complete suites of those advertised by the leading English and French seed houses, and collecting the principal kinds raised in this country. In the year 1888, I grew eighty-five different varieties and strains of cabbages and cauliflowers. I do not propose describing all these in this treatise or their comparative merits; of some of them I have yet something to learn, but I will endeavor to introduce with my description such notes as I think will prove of value to my fellow farmers and market gardeners.

I will here say in general of the class of early cabbages, that most of them have elongated heads between ovoid and conical in form. They appear to lack in this country the sweetness and tenderness that characterize some varieties of our drumhead, and, consequently, in the North when the drumhead enters the market there is but a limited call for them.

It may be well here to note a fundamental distinction between the drumhead cabbage of England and those of this country. In England the drumhead class are almost wholly raised to feed to stock. I venture the conjecture that owing in part, or principally, to the fact European gardeners have never had the motive, and, consequently, have never developed the full capacity of the drumhead as exampled by the fine varieties raised in this country. The securing of sorts reliable for heading being with them a matter of secondary consideration, seed is raised from stumps or any refuse heads that may be standing when spring comes round. For this reason English drumhead cabbage seed is better suited to raise a mass of leaves than heads, and always disappoints our American farmers who buy it because it is cheap with the expectation of raising cabbage for market. English-grown drumhead cabbage seed is utterly worthless for use in this country except to raise greens or collards.

The following are foreign varieties that are accepted in this country as standards, and for years have been more or less extensively cultivated: EARLY YORK, EARLY OXHEART, EARLY WINNIGSTADT, RED DUTCH, RED DRUMHEAD.

In my experience as a seed dealer, the Sugar Loaf and Oxheart are losing ground in the farming community, the Early Jersey Wakefield having, to a large extent, replaced them.

~Early York.~ Heads nearly ovoid, rather soft, with few waste leaves surrounding them, which are of a bright green color. Reliable for heading. Stump rather short. Plant two feet by eighteen inches. This cabbage has been cultivated in England over a hundred years. LITTLE PIXIE with me is earlier than Early York, as reliable for heading, heads much harder, and is of better flavor; the heads do not grow quite as large.

~Early Oxheart.~ Heads nearly egg-shaped, small, hard, few waste leaves, stumps short. A little later than Early York. Have the rows two feet apart, and the plants eighteen inches apart in the row.

~Early Winnigstadt.~ (A German cabbage.) Heads nearly conical in shape, having usually a twist of leaf at the top; larger than Oxheart, are harder than any of the early oblong heading cabbages; stumps middling short. Matures about ten days later than Early York. The Winnigstadt is remarkably reliable for heading, being not excelled in this respect when the seed has been raised with care, by any cabbage grown. It is a capital sort for early market outside our large cities, where the very early kinds are not so eagerly craved. It is so reliable for heading, that it will often make fine heads where other sorts fail; and I would advise all who have not succeeded in their efforts to grow cabbage, to try this before giving up their attempts. It is raised by some for winter use, and where the drumheads are not so successfully raised, I would advise my farmer friends to try the Winnigstadt, as the heads are so hard that they keep without much waste. Have rows two feet apart, and plant twenty inches to two feet apart in the rows.

~Red Dutch.~ Heads nearly conical, medium sized, hard, of a very deep red; outer leaves numerous, and not so red as the head, being somewhat mixed with green; stump rather long. This cabbage is usually planted too late; it requires nearly the whole season to mature. It is used for pickling, or cut up fine as a salad, served with vinegar and pepper. This is a very tender cabbage, and, were it not for its color, would be an excellent sort to boil; to those who have a mind to eat it with their eyes shut, this objection will not apply.

~Red Drumhead.~ Like the preceding, with the exception that the heads grow round, or nearly so, are harder, and of double the size. It is very difficult to raise seed from this cabbage in this country. I am acquainted with five trials, made in as many different years, two of which I made myself, and all were nearly utter failures, the yield, when the hardest heads were selected, being at about the rate of two great spoonfuls of seed from every twenty cabbages. French seed-growers are more successful, otherwise this seed would have to sell at a far higher figure in the market than any other sort.

~The Little Pixie.~ has much to recommend it, in earliness, quality, reliability for heading, and hardness of the head; earlier than Early York, though somewhat smaller.

Among those that deserve to be heartily welcomed and grow in favor, are the EARLY ULM SAVOY (for engraving and description of which see under head of Savoy), and the ST. DENNIS DRUMHEAD, a late, short-stumped sort, setting a large, round, very solid head, as large, but harder, than Premium Flat Dutch. The leaves are of a bluish-green, and thicker than those of most varieties of drumhead. Our brethren in Canada think highly of this cabbage, and if we want to try a new drumhead, I will speak a good word for this one.

~Early Schweinfurt~, or ~Schweinfurt Quintal~, is an excellent early drumhead for family use; the heads range in size from ten to eighteen inches in diameter, varying with the conditions of cultivation more than any other cabbage I am acquainted with. They are flattish round, weigh from three to nine pounds when well grown, are very symmetrical in shape, standing apart from the surrounding leaves. They are not solid, though they have the finished appearance that solidity gives; they are remarkably tender, as though blanched, and of very fine flavor. It is among the earliest of drumheads, maturing at about the same time as the Early Winnigstadt. As an early drumhead for the family garden, it has no superior; and where the market is near, and does not insist that a cabbage head must be hard to be good, it has proved a very profitable market sort.

The following are either already standard American varieties of cabbage, or such as are likely soon to become so; very possibly there are two or three other varieties or strains that deserve to be included in the list. I give all that have proved to be first class in my locality: EARLY WAKEFIELD, EARLY WYMAN, EARLY SUMMER, ALL SEASONS, HARD HEADING, SUCCESSION, WARREN, VANDERGAW, PEERLESS, NEWARK, FLAT DUTCH, PREMIUM FLAT DUTCH, STONE MASON, LARGE LATE DRUMHEAD, MARBLEHEAD MAMMOTH DRUMHEAD, AMERICAN GREEN GLAZED, FOTTLER'S DRUMHEAD, BERGEN DRUMHEAD, DRUMHEAD SAVOY, and AMERICAN GREEN GLOBE SAVOY. All of these varieties, as I have previously stated, are but improvements of foreign kinds; but they are so far improved through years of careful selection and cultivation, that, as a rule, they appear quite distinct from the originals when grown side by side with them, and this distinction is more or less recognized, in both English and American catalogues, by the adjective "American" or "English" being added after varieties bearing the same name.

~Early Wakefield~, sometimes called ~Early Jersey Wakefield.~ Heads mostly nearly conical in shape but sometimes nearly round, of good size for early, very reliable for heading; stumps short. A very popular early cabbage in the markets of Boston and New York. Plant two and a half feet by two feet. There are two strains of this cabbage, one a little later and larger than the other.

~Early Wyman.~ This cabbage is named after Capt. Wyman, of Cambridge, the originator. Like Early Wakefield the heads are usually somewhat conical, but sometimes nearly round; in structure they are compact. In earliness it ranks about with the Early Wakefield, and making heads of double the size, it has a high value as an early cabbage. Capt. Wyman had entire control of this cabbage until within the past few years, and, consequently, has held Boston Market in his own hands, to the chagrin of his fellow market gardeners, raising some seasons as many as thirty thousand heads. Have the rows from two to two and a half feet apart, and the plants from twenty to twenty-four inches apart in the row. Crane's Early is a cross between the Wyman and Wakefield, intermediate in size and earliness.

~Premium Flat Dutch.~ Large, late variety; heads either round or flat, on the top (varying with different strains); rather hard; color bluish green; leaves around heads rather numerous; towards the close of the season, the edge of some of the exterior leaves and the top of the heads assume a purple cast. The edges of the exterior leaves, and of the two or three that make the outside of the head, are quite ruffled, so that when grown side by side with Stone Mason, this distinction between the habit of growth of the two varieties is noticeable at quite a distance. Stumps short; reliable for heading. Have the rows three feet apart, and the plants from two and a half to three feet apart in the rows. This cabbage is very widely cultivated, and, in many respects, is an excellent sort to raise for late marketing. There are several strains of it catalogued by different seedsmen under various names, such as Sure Head, &c.

~Stone Mason.~ An improvement on the Mason, which cabbage was selected by Mr. John Mason of Marblehead, from a number of varieties of cabbage that came from a lot of seed purchased and planted as Savoys. Mr. John Stone afterwards improved upon the Mason cabbage, by increasing the size of the heads. Different growers differ in their standard of a Stone Mason cabbage, in earliness and lateness, and in the size, form, and hardness of the head. But all these varieties agree in the characteristics of being very reliable for heading, in having heads which are large, very hard, very tender, rich and sweet; short stumps, and few waste leaves. The color of the leaves varies from a bluish green to a pea-green, and the structure from nearly smooth to much blistered. In their color and blistering some specimens have almost a Savoy cast. The heads of the best varieties of Stone Mason range in weight from six to twenty-five pounds, the difference turning mostly on soil, manure, and cultivation.

The Stone Mason is an earlier cabbage than Premium Flat Dutch, has fewer waste leaves, and side by side, under high cultivation, grows to an equal or larger size, while it makes heads that are decidedly harder and sweeter. These cabbages are equally reliable for heading. I am inclined to the opinion that under poor cultivation the Premium Flat Dutch will do somewhat better than the Stone Mason.

Until the introduction of Fottler's Drumhead it was the standard drumhead cabbage in the markets of Boston and other large cities of the North. Have the rows three feet apart, and the plants from two to three feet apart in the row.

~Large Late Drumhead.~ Heads large, round, sometimes flattened at the top, close and firm; loose leaves numerous; stems short; reliable for heading, hardy, and a good keeper. The name "Large Late Drumhead" includes varieties raised by several seedsmen in this country, all of which resemble each other in the above characteristics, and differ in but minor points. Have rows three feet apart, and plants from two and a half to three feet apart in the row.

~Marblehead Mammoth Drumhead.~ This is the largest of the cabbage family, having sometimes been grown to weigh over ninety pounds to the plant. It originated in Marblehead, Mass., being produced by Mr. Alley, probably from the Mason, by years of high cultivation and careful selection of seed stock. I introduced this cabbage and the Stone Mason to the general public many years ago, and it has been pretty thoroughly disseminated throughout the United States. Heads varying in shape between hemispherical and spherical, with but few waste leaves surrounding them; size very large, varying from fifteen to twenty inches in diameter, and, in some specimens, they have grown to the extraordinary dimensions of twenty-four inches. In good soil, and with the highest culture, this variety has attained an average weight of thirty pounds by the acre. Quality, when well grown, remarkably sweet and tender, as would be inferred from the rapidity of its growth. Cultivate in rows four feet apart, and allow four feet between the plants in the rows. Sixty tons of this variety have been raised from a single acre.

~American Green Glazed.~ Heads loose, though rather large, with a great body of waste leaves surrounding them; quality poor; late; stump long. This cabbage was readily distinguished among all the varieties in my experimental plot by the deep, rich green of the leaves, with their bright lustre as though varnished. It is grown somewhat extensively in the South, as it is believed not to be so liable to injury from insects as other varieties. Plant two and a half feet apart each way. I would advise my Southern friends to try the merits of other kinds before adopting this poor affair. I know, through my correspondence, that the Mammoth has done well as far South as Louisiana and Cuba, and the Fottler, in many sections of the South, has given great satisfaction.

~Fottler's Early Drumhead.~ Several years ago a Boston seedsman imported a lot of cabbage seed from Europe, under the name of Early Brunswick Short Stemmed. It proved to be a large heading and very early Drumhead. The heads were from eight to eighteen inches in diameter nearly flat, hard, sweet, and tender in quality; few waste leaves; stump short. In earliness it was about a fortnight ahead of the Stone Mason. It was so much liked by the market gardeners that the next season he ordered a larger quantity; but the second importation, though ordered and sent under the same name, proved to be a different and inferior kind, and the same result followed one or two other importations. The two gardeners who received seed of the first importation brought to market a fine, large Drumhead, ten days or a fortnight ahead of their fellows. The seed of the true stock was eagerly bought up by the Boston market gardeners, most of it at five dollars an ounce. After an extensive trial on a large scale by the market farmers around Boston, and by farmers in various parts of the United States, Fottler's Cabbage has given great satisfaction, and become a universal favorite, and when once known it, and especially the improved strain of it, known as Deep Head, is fast replacing some of the old varieties of Drumhead. Very reliable for heading.

~Vandergaw Cabbage.~ This new Long Island Cabbage must be classed as A No. 1 for the midsummer and late market. It is as sure to head as the

Succession, and has some excellent characteristics in common.

It makes large, green heads, hard, tender, and crisp. This is an acquisition.

~The Warren Cabbage.~ This first-class cabbage is closely allied to, but an improvement on, the old Mason Cabbage of twenty-five years ago. It makes a head deep, round, and very hard, the outer leaves wrapping it over very handsomely. In reliability for heading no cabbage surpasses it; a field of them when in their prime is as pretty a sight as a cabbage man would wish to see. It comes in as early as some strains of Fottler, and a little earlier than others. A capital sort to succeed the Early Summer. The heads being very thick through, and nearly round, make it an excellent sort to carry through the winter, as it "peels" well, as cabbage-growers say. Ten inches in diameter, in size it is just about right for profitable marketing. A capital sort, exceedingly popular among market-man in this vicinity.

~Early Bleichfeld Cabbage.~ I find the Bleichfeld to be among the earliest of the large, hard-heading Drumheads, maturing earlier than the Fottler's Brunswick. The heads are large, very solid, tender when cooked, and of excellent flavor. The color is a lighter green than most varieties and it is as reliable for heading as any cabbage I have ever grown. The above engraving I have had made from a photograph of a specimen grown on my grounds.

~Danish Drumhead Cabbage.~ In 1879, Mr. Edward Abelgoord wrote me from Canada, that he raised a large Drumhead Cabbage, the seed of which was brought from Denmark, which was the best kind of cabbage that he had seen in that latitude (46?, being very valuable for the extreme North. It was earlier than Fottler's Drumhead, and made large, flat heads, of excellent flavor, and was so reliable for heading. I raised a field of this new cabbage, and it proved a large, flat, early Drumhead, very reliable for heading.

~The Reynolds Early Cabbage.~ In the year 1875, Mr. Franklin Reynolds, of this town, crossed the Cannon-Ball Cabbage on the Schweinfurt Quintal, by carefully transferring the pollen of the former on the latter, the stamens having first been removed, and immediately tying muslin around the impregnated blossoms to keep away all insects. The results were a few ripe seeds. These were carefully saved and planted the next season, when the product showed the characteristics of the two parents. The best heads were selected from the lot, and, from these, seeds were raised. Several selections were made of the choicest heads from year to year; and I now have the pleasure of introducing the results, a new cabbage which combines the good qualities of both its parents.

The flavor of this new cabbage is rich, tender, and sweet, being superior to the general Drumhead class, making it a very superior variety for family use, and also for marketing when there is not a long transportation. None of the scores of varieties I have ever grown has a shorter stump than this; the heads appear to rest directly on the ground, and no one is surer to head.

~All-Seasons Cabbage.~ This new cabbage is the result of a cross made by a Long Island gardener between the Flat Dutch and a variety of Drumhead. The result is a remarkably large, early Drumhead, that matures close in time with the Early Summer, while it is from one third to one half larger. It is an excellent variety either as an early or late sort; the roundness of the head, leaving a thick, solid cabbage, should it become necessary, as is often the case with those marketed in the spring, to peel off the outer layer of leaves. Heads large in size, solid and tender, and rich flavored when cooked. It has already, in three years, verified the prophecy I made when sending it out, and become a standard variety in some localities.

~Gregory's Hard-Heading Cabbage.~ I am not acquainted with any variety of cabbage (I believe I have raised about all the native and foreign varieties that have been catalogued) that makes so hard a head as does the "Hard-heading" when fully matured. Neither am I acquainted with any variety that is so late a keeper as is this; the German gardener, from whom I obtained it, said that it gave him, and his friends who had it, complete control of the Chicago market for about a fortnight after all other varieties had "played out." My own experience with it tends to confirm this statement, for under the same conditions it kept decidedly later than all my other varieties, was greener in color, and when planted out they were so late to push seed-shoots that I almost despaired of getting a crop of seed. I find, also, that they are much less inclined to burst than any of the hard-heading varieties. Heads grow to a good market size, are more globular than Flat Dutch; and, as might be presumed, of great weight in proportion to their size. The color is a peculiar green, rather more of an olive than most kinds of cabbage. About a fortnight later than Flat Dutch. For late fall, winter, and spring sales plant 3 by 3 the first of June.

~Early Deep-Head Cabbage.~ This is a valuable improvement on the Fottler made by years of careful selection and high cultivation by Mr. Alley of Marblehead, a famous cabbage grower, who, as the name indicates, has produced a deeper, rounder heading variety than the original Fottler, thus making what that was not, an excellent sort for winter and spring marketing. It has all the excellent traits of its parent in reliability for making large, handsome heads.

~Bergen Drumhead.~ Heads round, rather flat on the top, solid; leaves stout, thick, and rather numerous; stump short. With me, under same cultivation, it is later than Stone Mason. It is tender and of good flavor. A popular sort in many sections, particularly in the markets of New York City. Have the plants three feet apart each way.

SAVOY CABBAGES.

The Savoys are the tenderest and richest-flavored of cabbages, though not always as sweet as a well-grown Stone Mason; nor is a Savoy grown on poor soil, or one that has been pinched by drouth, as tender as a Stone Mason that has been grown under favoring circumstances; yet it remains, as a rule, that the Savoy surpasses all other cabbages in tenderness, and in a rich, marrow-like flavor. The Savoys are also the hardiest of the cabbage tribe, enduring in the open field a temperature within sixteen degrees of zero without serious injury; and if the heads are not very hard they will continue to withstand repeated changes from freezing to thawing for a couple of months, as far north as the latitude of Boston. A degree of freezing improves them, and it is common in that latitude to let such as are intended for early winter use, in the family, remain standing in the open ground where they grew, cutting the heads as they are wanted.

As a rule Savoys neither head as readily (the "Improved American Savoy" being an exception) nor do the heads grow as large as the Drumhead varieties; indeed, most of the kinds in cultivation are so unreliable in these respects as to be utterly worthless for market purposes, and nearly so for the kitchen garden.

~The Drumhead Savoy.~ This, as the name implies, is the result of a cross between a Savoy and a Drumhead cabbage, partaking of the characteristics of each. Many of the cabbages sold in the market as Savoy are really this variety. One variety in my experimental garden, which I received as TOUR'S SAVOY (evidently a Drumhead variety of the Savoy), proved to be much like Early Schweinfurt in earliness and style of heading; the heads were very large, but quite loose in structure; I should think it would prove valuable for family use.

It is a fact that does not appear to be generally known that we have among the Savoys some remarkably early sorts which rank with the earliest varieties of cabbage grown. Pancalier and Early Ulm Savoy are earlier than that old standard of earliness, Early York; Pancalier being somewhat earlier than Ulm.

~Pancalier~ is characterized by very coarsely blistered leaves of the darkest-green color; the heads usually gather together, being the only exception I know of to the rule that cabbage heads are made up of overlapping leaves, wrapped closely together. It has a short stump, and with high cultivation is reliable for heading. The leaves nearest the head, though not forming a part of it, are quite tender, and may be cooked with the head. Plant fifteen by thirty inches.

~Early Ulm Savoy~ is a few days later than Pancalier, and makes a larger head; the leaves are of a lighter green and not so coarsely blistered; stump short; head round; very reliable for heading. It has a capital characteristic in not being so liable as most varieties to burst the head and push the seed shoot immediately after the head is matured. For first early, I know no cabbages so desirable as these for the kitchen garden.

The ~Early Dwarf Savoy~ is a desirable variety of second early. The heads are rather flat in shape, and grow to a fair size. Stumps short; reliable for heading.

~Improved American Savoy.~ Everything considered, this is the Savoy, "par excellence," for the market garden. It is a true Savoy, the heads grow to a large size, from six to ten inches in diameter, varying, of course, with soil, manure, and cultivation. In shape the heads are mostly globular, occasionally oblong, having but few waste leaves, and grow very solid. Stump short. In reliability for heading it is unsurpassed by any other cabbage.

~Golden Savoy~ differs from other varieties in the color of the head, which rises from the body of light green leaves, of a singular pale yellow color, as though blanched. The stumps are long, and the head rather small, a portion of these growing pointed. It is very late, not worth cultivating, except as a curiosity.

~Norwegian Savoy.~ This is a singular half cabbage, half kale--at least, so it has proved under my cultivation. The leaves are long, narrow, tasselated, and somewhat blistered. The whole appearance is very singular and rather ornamental. I have tried this cabbage twice, but have never got beyond the possible promise of a head.

~Victoria Savoy~, ~Russian Savoy~, and ~Cape Savoy~, tested in my experimental garden, did not prove desirable either for family use or for market purposes.

~Feather Stemmed Savoy.~ This is a cross between the Savoy and Brussels sprouts, having the habit of growth of Brussels sprouts.

OTHER VARIETIES OF CABBAGE.

I will add notes on some other varieties which have been tested, from year to year, in my experimental plot. The results from tests of different strains of standard sorts, I have not thought it worth the while to record.

~Cannon Ball.~ The heads are usually spherical, attaining to a diameter of from five to nine inches, with the surrounding leaves gathered rather closely around them; in hardness and relative weight it is excelled by but few varieties. Stump short. It delights in the highest cultivation possible. It is about a week later than Early York. In those markets where cabbages are sold by weight, it will pay to grow for market; it is a good cabbage for the family garden.

~Early Cone~, of the Wakefield class, but with me not as early.

~Garfield Pickling~, of late variety, of the conical class.

~Cardinal Red.~ A large, late variety of red; but on my grounds, it is not equal to Red Drumhead.

~Vilmorin's Early Flat Dutch.~ Not quite as large as Early Summer, though about as early and resembles it in shape of head.

~Royal German Drumhead.~ Reliable for heading.

~Large White Solid Magdeburg.~ A late Drumhead; short stumped; reliable for heading. Medium late.

~Pak Choi.~ Evidently of the Kale class; no heads.

~Chou de Burghlez~ and ~Chou de Milan~. These are coarse, loose, small heading varieties, allied to Kale. The latter is of the Savoy class.

~Earliest Erfurt Blood-Red.~ Decidedly the earliest of the red cabbages. Very reliable for heading. A Drumhead; smaller than Red Drumhead. Very dark red.

~Empress.~ Resembles Wyman in size and shape; but the heads are more pointed, and it makes head earlier. Heads well.

~Schlitzer.~ This makes heads mostly shaped like the Winnigstadt, but a third larger. Its mottling of green and purple gives it a striking appearance. Early and very reliable for heading. Heads are not very hard; but, when cooked, are just about as tender and rich-flavored as the Savoy. Promises to be an excellent sort for family use.

~Rothelburg.~ An early sure heading variety of the Drumhead class. Heads of medium size; resembling in shape Deep Head.

~Sure Head.~ A strain of Flat Dutch. A late variety; heads deeper than Fottler, but with me not so reliable.

~Dark Red Pointed.~ Resembles Winnigstadt in shape. About as late as Red Dutch, and not as desirable.

~Bacalan Late.~ In shape resembles Winnigstadt. Grow a little wild.

~Amack.~ A late variety. Heads generally nearly globular and quite hard. Very reliable for heading.

~Bangholm.~ First of all. As early as the earliest, but very small,--not as large as Little Pixie.

~Early Enfield Market.~

~Tourleville.~ Heads resemble Wakefield in form; but, with me, are neither so large nor so large, and are more inclined to burst.

~Danish Round Winter.~ A late variety; bearing deep, hard heads on long stumps.

~Dwarf Danish.~ Late. Reliable to head; uneven in time of heading. Worth planting for market.

~Danish Ball Drumhead.~ Heads not characterized by globular shape, but rather flattish. Irregular in length of stump.

~Early Paris.~ Closely resembles Wakefield.

~Very Early Etampes.~ Earlier than Wakefield. Shape partakes of both Oxheart and Wakefield.

~Early Mohawk.~ Light green in color; a good header, but not so hard heading as Fottler. Appears to have a little of the Savoy cross in it.

~Sure Head.~ A late variety of the Dutch class; reliable for heading; stump rather long.

~Excelsior.~ A variety which is of the Fottler class, but makes smaller sized heads.

~Louisville Drumhead.~ Of the flat Dutch type; nearly as early as Early Summer.

~Early Advance.~ Of the Wakefield type. With me it is full as early as Wakefield, and considerably larger. Rather coarser in structure.

~Market Garden.~ Of the Fottler class; very reliable for heading. Heads of good size, but rather coarser than the Deep Head.

~Chase's Excelsior.~ A second early; much like Fottler; heads finely.

~Bloomsdale Early Market.~ With me this is not as good a variety as Wakefield.

~Berkshire Beauty.~ There appear to be fine possibilities in this cabbage, which have not yet been developed into uniformity.

~Landredth's Extra Early.~ With me it does not prove as early as Wakefield, and does not head as well.

~Bridgeport Late Drumhead.~ A large Drumhead; in size, between Stone Mason and Marblehead Mammoth. Reliable for heading, but does not head as hard as either of these varieties. Not inclined to burst.

~Large French Oxheart~ closely resembles Early Oxheart, but grows to double the size, and is about ten days later; quality usually good.

~Early Sugar Loaf.~ Heads shaped much like a loaf of sugar standing on its smaller end, resembling, as Burr well says, a head of Cos lettuce in its shape, and in the peculiar clasping of the leaves about the head. Heads rather hard, medium size; early, and tender. It is said not to stand the heat as well as most sorts.

~Large Brunswick Short-Stemmed.~ (English seed.) Late, long-stumped, wild, plenty of leaves, almost no head; bears but a slight resemblance to Fottler's Drumhead.

~Early Empress.~ Cabbages well; heads conical; early.

~Robinson's Champion Ox Drumhead.~ Stump long; heads soft and not very large; wild.

~English Winnigstadt.~ Long-stumped; irregular; not to be compared with French stock.

~Blenheim.~ Early; heads mostly conical; of good size.

~Shillings Queen.~ Early; heads conical; stumps long.

~Carter's Superfine Early Dwarf.~ Surpasses in earliness and hardness of head. Closely allied to Little Pixie.

~Enfield Market Improved.~ Most of the heads were flat; rather wild; not to be compared with Fottler.

~Kemp's Incomparable.~ Long-headed; heads, when mature, do not appear to burst as readily as with most of the conical class.

~Fielderkraut.~ Closely resembles Winnigstadt, with larger and longer heads and stump; requires more room than Winnigstadt.

~Ramsay's Winter Drumhead.~ Closely resembles St. Dennis. I think it is the same.

~Pomeranian Cabbage.~ Heads very long; quite large for a conical heading sort; very symmetrical and hard; color, yellowish-green. It handles well, and I should think would prove a good keeper. Medium early.

~Alsacian Drumhead.~ Stump long; late; wild.

~Marbled Bourgogne.~ Stumps long; heads small and hard; color, a mixture of green and red.

CABBAGE GREENS.

In the vicinity of our large cities, the market gardeners sow large areas very thickly with cabbage seed, early in the spring, to raise young plants to be sold as greens. The seed is sown broadcast at the rate of ten pounds and upwards to the acre. Seed of the Savoy cabbage is usually sown for this purpose, which may be sometimes purchased at a discount, owing to some defect in quality or purity, that would render it worthless for planting for a crop of heading cabbage.

The young plants are cut off about even with the ground, when four or five inches high, washed, and carried to market in barrels or bushel boxes. The price varies with the state of the market, from 12 cents to $3 a barrel, the average price in Boston market being about a dollar. With the return of spring most families have some cabbage stumps remaining in the cellar; these can be planted about a foot apart in some handy spot along the edge of the garden, where they will not interfere with the general crop, setting them under ground from a quarter to a half their length, depending on the length of the stumps. They will soon be covered with green shoots, which should be used as greens before the blossom buds show themselves, as they then become too strong to be agreeable. If the spot is rich and has been well dug, the rapidity of growth is surprising; and if the shoots are frequently gathered, many nice messes of greens can be grown from a few stumps. Farmers in Northern Vermont tell me, that if they break off each seed shoot as soon as it shows itself, close home to the stump, nice little heads will push out on almost every stump. In England, where the winter climate is much milder than that of New England, it is the practice to raise a second crop of heads in this way. In my own neighborhood I have seen an acre from which a crop of drumhead cabbage had been cut off early in the season, every stump on which had from three to six hard heads, varying from the size of a hen's egg to that of a goose egg; but to get this second growth of heads, as much of the stump and leaves should be left as possible, when cutting out the original head. As in the cabbage districts of the North little or no use is made of this prolific after growth, it is worse than useless to suffer the ground to be exhausted by it; the stump should be pulled by the potato hoe as soon as the heads are marketed. When cabbages are planted out for seed, if, for any reason, the seed shoot fails to push out, and at times when it does push out, fine sprouts for greens will start below the head; when the stock of these sprouts becomes too tough for use, the large leaves may be stripped from them and cooked. I usually break off the tender tops of large sprouts, and then strip off the tenderest of the large leaves below.

CABBAGE FOR STOCK.

No vegetable raised in the temperate zone, Mangold Wurtzel alone excepted, will produce as much food to the acre, both for man and beast, as the cabbage. I have seen acres of the Marblehead Mammoth drumhead which would average thirty pounds to each cabbage, some specimens weighing over sixty pounds. The plants were four feet apart each way which would give a product of over forty tons to the acre; and I have tested a crop of Fottler's that yielded thirty tons of green food to the half acre. Other vegetables are at times raised for cattle feed, such as potatoes, carrots, ruta bagas, mangold wurtzels; a crop of potatoes yielding four hundred bushels to the acre at sixty pounds the bushel would weigh twelve tons; a crop of carrot yielding twelve hundred bushels to the acre would weigh thirty tons; ruta bagas sometimes yield thirty tons; and mangolds as high as seventy tons to the acre. I have set all these crops at a high capacity for fodder purposes; the same favoring conditions of soil, manure, and cultivation that would produce four hundred bushels of potatoes, twelve hundred bushels of carrots, and thirty-five tons of ruta baga turnips, would give a crop of forty tons of the largest variety of drumhead cabbage. If we now consider the comparative merits of these crops for nutriment, we find that the cabbage excels them all in this department also. The potatoes abound in starch, the mangold and carrot are largely composed of water, while the cabbage abounds in rich, nitrogeneous food.

Prof. Stewart states that cabbage for milch cows has about the same feeding value as sweet corn ensilage, and makes the value not over $3.40 per ton. Now it is admitted by general current that the value of common ensilage, which is inferior to that made from sweet corn, is, when compared with good English hay, as 3 to 1. This would make cabbages for milch cows worth not far from $7.00 per ton.

When cabbage is kept for stock feed later than the first severe frost, if the quantity is large there is considerable waste even with the best of care. The loose leaves should be fed first, and the heads kept in a cool place, not more than two or three deep, at as near the freezing point as possible. If it has been necessary to cut the heads from the stumps, they may be piled, after the weather has set in decidedly cold, conveniently near the barn, and kept covered with a foot of straw or old litter. As long as a cabbage is kept frozen there is no waste to it; but if it be allowed to freeze and thaw two or three times, it will soon rot with an awful stench. I suspect that it is this rotten portion of the cabbage that often gives the bad flavor to milk. On the other hand, if it is kept in too warm and dry a place, the outer leaves will dry, turning yellow, and the whole head lose in weight,--if it be not very hard, shrivelling, and, if hard, shrinking. If they are kept in too warm and wet a place, the heads will decay fast, in a black, soft rot. The best way to preserve cabbages for stock into the winter, is to place them in trenches a few inches below the surface, and there cover with from a foot to two feet of coarse hay or straw, the depth depending on the coldness of the locality. When the ground has been frozen too hard to open with a plough or spade, I have kept them until spring by piling them loosely, hay-stack shape, about four feet high, letting the frost strike through them, and afterwards covering with a couple of feet of eel-grass; straw or coarse hay would doubtless do as well.

I have treated of cabbage thus far when grown specially for stock; in every piece of cabbage handled for market purposes, there is a large proportion of waste suitable for stock feed, which includes the outside leaves and such heads as have not hardened up sufficiently for market. On walking over a piece just after my cabbages for seed stock have been taken off, I note that the refuse leaves that were stripped from the heads before pulling are so abundant they nearly cover the ground. If leaves so stripped remain exposed to frost, they soon spoil; or, if earlier in the season they are exposed to the sun, they soon become yellow, dry, and of but little value. They can be rapidly collected with a hay fork and carted, if there be but a few, into the barn; should there be a large quantity, dump them within a convenient distance of the barn or feeding ground, but not where the cattle can trample them, and spread them so that they will be but a few inches in depth. If piled in heaps they will quickly heat; but even then, if not too much decayed, cattle will eat them with avidity. Cabbages are hardy plants, and loose heads will stand a good deal of freezing and thawing without serious injury. They are not generally injured with the thermometer 16?below freezing. The waste, after the seed and all market cabbage are removed, brings me about $10 per acre on the ground, for cow feed.

If cabbage is fed to cows in milk without some care, it will be apt to give the milk a strong cabbage flavor; all the feed for the day should be given early in the morning. Beginning with a small quantity, and gradually increasing it, the dairy man will soon learn his limits. The effect of a liberal feed to milk stock is to largely increase the flow of milk. Avoid feeding to any extent while the leaves are frozen.

An English writer says: "The cabbage comes into use when other things begin to fail, and it is by far the best succulent vegetable for milking cows,-- keeping up the yield of milk, and preserving, better than any other food, some portion of the quality which cheese loses when the cows quit their natural pasturage. Cows fed on cabbages are always quiet and satisfied, while on turnips they often scour and are restless. When frosted, they are liable to produce hoven, unless kept in a warm shed to thaw before being used; fifty-six pounds given, at two meals, are as much as a large cow should have in a day. Frequent cases of abortion are caused by an over-supply of green food. Cabbages are excellent for young animals, keeping them in health, and preventing 'black leg.' A calf of seven months may have twenty pounds a day."

RAISING CABBAGE SEED.

Cabbage seed in England, particularly of the drumhead sorts, is mostly raised from stumps, or from the refuse that remains after all that is salable has been disposed of. The agent of one of the largest English seed houses, a few years since, laughed at my "wastefulness," as he termed it, in raising seed from solid heads. In our country, cabbage seed is mostly raised from soft, half-formed heads, which are grown as a late crop, few, if any of them, being hard enough to be of any value in the market. Seedsmen practise selecting a few fine, hard heads, from which to raise their seed stock. It has been my practice to grow seed from none but extra fine heads, better than the average of those carried to market. I do this on the theory that no cabbage can be too good for a seedhead, if the design is to keep the stock first-class. Perhaps such strictness may not be necessary; but I had rather err in setting out too good heads than too poor ones; besides, the great hardness obtained by the heads of the Stone Mason, makes it possible, at least, that I am right. Cabbage raised from seed grown from stumps are apt to be unreliable for heading, and to grow long-stumped, though under unfavorable conditions, long-stumped and poor-headed cabbage may grow from the best of seed. To have the best of seed, all shoots that start below the head should be broken off. To prevent the plants falling over after the seed-stalks are grown, dig deep holes, and plant the entire stump in the ground. Scarecrows should be set up, or some like precaution be taken, to keep away the little seed-birds, that begin to crack the pods as soon as they commence to ripen. A plaster cat is a very good scarecrow to frighten away birds from seed and small fruits, if its location is changed every few days.

I find that the pods of cabbage seed grown South are tough, and not brittle, like those grown North, and hence that they are injured but little, if any, by seed birds. When the seed-pods have passed what seedsmen call their "red" stage, they begin to harden; as soon as a third of them are brown, the entire stalk may be cut and hung up in a dry, airy place, for a few days, when the seed will be ready for rubbing or threshing out. Different varieties should be raised far apart to insure purity; and cabbage seed had better not be raised in the vicinity of turnip seed. There is some difference of opinion as to the effect of growing these near each other; where the two vegetables blossom at the same time, I should fear an admixture. When the care requisite to select good seed stock, and the trouble, and, often, great loss, in keeping it over winter, planting it in isolated locations, protecting it from wind and weather, guarding it from injury from birds and other enemies, gathering it, cleaning it, are all considered, few men will find that they can afford to raise their own seed, provided they can buy it from reliable seedsmen.

COOKING CABBAGE, SOUR-KROUT, ETC.

Cabbage when boiled with salt pork, as it is mostly used, is the food for strong and healthy digestive powers; but when eaten in its raw state, served with vinegar and pepper, it is considered one of the most easily digested articles of diet. In the process of cooking, even with the greatest care, a large portion of the sweetness is lost. The length of time required to cook cabbage by boiling varies with the quality, those of the best quality requiring about twenty minutes, while others require an hour. In cooking put it into boiling water in which a little salt and soda has been sprinkled, which will tend to preserve the natural green color. It will be well to change the water once. The peculiar aroma given out by cabbage when cooking is thought to depend somewhat on the manner in which it is grown; those having been raised with the least rank manure having the least. I think this is one of the whims of the community. By using some varieties of boilers all steam is carried into the fire, and there is no smell in the house.

To Pickle, select hard heads, quarter them, soak in salt and water four or five days, then drain and treat as for other pickles, with vinegar spiced to suit.

For Cold Slaw, select hard heads, halve and then slice up these halves exceedingly fine. Lay these in a deep dish, and pour over vinegar that has been raised to the boiling point in which has been mixed a little pepper and salt.

Sour-Krout. Take large, hard-headed drumheads, halve, and cut very fine; then pack in a clean, tight barrel, beginning with a sprinkling of salt, and following with a layer of cabbage, and thus alternating until the barrel is filled. Now compact the mass as much as possible by pounding, after which put on a well-fitting cover resting on the cabbage, and lay heavy weights or a stone on this. When fermented it is ready for use. To prepare for the table fry in butter or fat.

The outer green leaves of cabbages are sometimes used to line a brass or copper kettle in which pickles are made in the belief that the vinegar extracts the coloring substance (chlorophyl) in the leaves, and the cucumbers absorbing this acquire a rich green color. Be not deceived by this transparent cheat, O simple housewife! the coloring matter comes almost wholly from the copper or brass behind those leaves; and, instead of an innocent vegetable pigment, your green cucumbers are dyed with the poisonous carbonate of copper.

CABBAGES UNDER GLASS.

The very early cabbages usually bringing high prices, the enterprising market gardener either winters the young plants under glass or starts them there, planting the seed under its protecting shelter long before the cold of winter is passed. When the design is to winter over fall grown plants, the seed are planted in the open ground about the middle of September, and at about the last of October they are ready to go into the cold frames, as such are called that depend wholly on the sun for heat. Select those having short stumps and transplant into the frames, about an inch and a half by two inches apart, setting them deep in the soil up to the lower leaves, shading them with a straw mat, or the like, for a few days, after which let them remain without any glass over them until the frost is severe enough to begin to freeze the ground, then place over the sashes; but bear in mind that the object is not to promote growth, but, as nearly as possible, to keep them in a dormant state, to keep them so cold that they will not grow, and just sufficiently protected to prevent injury from freezing. With this object in view the sashes must be raised whenever the temperature is above freezing, and this process will so harden the plants that they will receive no serious injury though the ground under the sash should freeze two inches deep; cabbage plants will stand a temperature of fifteen to twenty degrees below the freezing point. A covering of snow on the sash will do no harm, if it does not last longer than a week or ten days, in which case it must be removed. There is some danger to be feared from ground mice, who, when everything else is locked up by the frost, will instinctively take to the sash, and there cause much destruction among the plants unless these are occasionally examined. When March opens remove the sash when the temperature will allow, replacing it when the weather is unseasonably cold, particularly at night. The plants may be brought still farther forward by transferring them from the hot-bed when two or three inches high to cold frames, having first somewhat hardened them. When so transferred plant them about an inch apart, and shield from the sun for two or three days. After this they may be treated as in cold frames. The transfer tends to keep them stocky, increases the fibrous roots and makes the plants hardier. As the month advances it may be left entirely off, and about the first of April the plants may be set out in the open field, pressing fine earth firmly around the roots.

When cabbages are raised in hot-beds the seed, in the latitude of Boston, should be planted on the first of March; in that of New York, about a fortnight earlier. When two or three inches high, which will be in three or four weeks, they should be thinned to about four or less to an inch in the row. They should now be well hardened by partly drawing off the sashes in the warm part of the day, and covering at night; as the season advances remove the sashes entirely by day, covering only at night. By about the middle of April the plants will be ready for the open ground.

When raised in cold frames in the spring, the seed should be planted about the first of April, mats being used to retain by night the solar heat accumulated during the day. As the season advances the same process of hardening will be necessary as with those raised in hot-beds.

COLD FRAME AND HOT-BED.

To carry on hot-beds on a large scale successfully is almost an art in itself, and for fuller details I will refer my readers to works on gardening. Early plants, in a small way, may be raised in flower pots or boxes in a warm kitchen window. It is best, if practicable, to have but one plant in each pot, that they may grow short and stocky. If the seed are not planted earlier than April, for out-of-door cultivation, a cold frame will answer.

For a cold frame select the locality in the fall, choosing a warm location on a southern slope, protected by a fence or building on the north and north-west. Set posts in the ground, nail two boards to these parallel to each other, one about a foot in height, and the other towards the south about four inches narrower; this will give the sashes resting on them the right slope to shed the rain and receive as much heat as possible from the sun. Have these boards at a distance apart equal to the length of the sash, which may be any common window sash for a small bed, while three and a half feet is the length of a common gardener's sash. If common window sash is used cut channels in the cross-bars to let the water run off. Dig the ground thoroughly (it is best to cover it in the fall with litter, to keep the frost out) and rake out all stones or clods; then slide in the sash and let it remain closed for three or four days, that the soil may be warmed by the sun's rays. The two end boards and the bottom board should rise as high as the sash, to prevent the heat escaping, and the bottom board of a small frame should have a strip nailed inside to rest the sash on. Next rake in, thoroughly, guano, or phosphate, or finely pulverized hen manure, and plant in rows four to six inches apart. As the season advances raise the sashes an inch or two, in the middle of the day, and water freely, at evening, with water that is nearly of the temperature of the earth in the frame. As the heat of the season increases whitewash the glass, and keep them more and more open until just before the plants are set in open ground, then allow the glass to remain entirely off, both day and night, unless there should be a cold rain. This will harden them so that they will not be apt to be injured by the cabbage beetle, as well as chilled and put back by the change. Should the plants be getting too large before the season for transplanting, they should be checked by root pruning,--drawing a sharp knife within a couple of inches of the stalk. If it is desirable still further to check their growth, or harden them, transplant into another cold frame, allowing each plant double the distance it before occupied.

The structure and management of a hot-bed is much the same as that of a cold frame, with the exception that the sashes are usually longer and the back and front somewhat higher; being started earlier the requisite temperature has to be kept up by artificial means, fermenting manure being relied upon for the purpose; and the loss of this heat has to be checked more carefully by straw matting, and, in the far North, by shutters also. In constructing it, horse-manure, with plenty of litter, and about a quarter its bulk in leaves, if attainable, all having been well mixed together, is thrown into a pile, and left for a few days until steam escapes, when the mass is again thrown over and left for two or three days more, after which it is thrown into the pit (or it may be placed directly on the surface) which is lined with boards, from eighteen inches to two feet in depth, when it is beaten down with a fork and trodden well together. The sashes are now put on and kept there until heat is developed. The first intense heat must be allowed to pass off, which will be in about three days after the high temperature is reached. Now throw on six or eight inches of fine soil, in which mix well rotted manure, free from all straw, or rake in, thoroughly, superphosphate, or guano, at the rate of two thousand pounds to the acre, and plant the seed as in cold frame. Harden the plants as directed in preceding paragraph.

CAULIFLOWER, BROCCOLI, BRUSSELS-SPROUTS, KALE, AND SEA-KALE.

My treatise on the cabbage would hardly be complete without some allusion to such prominent members of the Brassica family as the cauliflower, broccoli, brussels-sprouts, and kale.

~Cauliflower.~ Wrote the great Dr. Johnson: "Of all the flowers of the garden, give me the cauliflower." Whether from this we are to infer the surpassing excellence of this member of the Brassica family, or that the distinguished lexicographer meant emphatically to state his preference of utility to beauty (perhaps our own Ben. Franklin took a leaf from him), each reader must be his own judge; but be that as it may, it remains true, beyond all controversy, that the cauliflower, in toothsome excellence, stands at the head of the great family of which it is a member. To be successful, and raise choice cauliflowers, is the height of the ambition of the market gardener; and, with all his experience, and with every facility at hand, he does not expect full success oftener than three years in four. The cauliflower, like the strawberry, is exceedingly sensitive to the presence or absence of sufficient water, and success or failure with the crop may turn on its having a full supply from the time they are half grown. The finest specimens raised in Europe are grown in beds, which are kept well watered from the supply which runs between them; and the most successful growers in the country irrigate their crops during periods of drouth. Cauliflowers do best on deep, rich, rather moist soils. In the way of food, they want the very best, and plenty of it at that. The successful competitor, who won the first prize at the great Bay State Fair, to the disgusted surprise of a grower justly famous for his almost uniform success in winning the laurels, whispered in my ear his secret: "R. manures very heavily in the spring for his crop. I manure very heavily both fall and spring." In manuring, therefore, do as well by them as by your heaviest crop of large drumhead cabbage, using rich and well-rotted manure, broadcast, with dissolved bone or ashes, or both, in the drill. Plough deep, and work the land very thoroughly, two ploughings, with a harrowing between, are better than one. Give plenty of room; three by three for the smaller sorts, and three by three and a half for the later and larger. They need the same cultivation, and, being subject to the same diseases and injury from insect enemies, need the same protection as their cousins of the cabbage tribe. In raising for the summer market, start in the cold frame, or plant as early as the ground can be worked, that the plants may get well started before the dry season, or the crop will be likely to make such small heads "buttons" as to be practically a failure. For late crop, plant seed in the hills where they are to grow, from the 20th of May to the middle of June. The crop ripens somewhat irregularly. When there is danger from frost, the later heads should be pulled and stored, with both roots and leaves, being crowded, standing as they grew, into a cold cellar or cold pit, when they will continue growing. As soon as the heads begin to form, they should be protected from sunlight by either half breaking off the outer leaves and bending them over them, or by gathering these leaves loosely together and confining them loosely by rough pegs, or by tying them together with a wisp of rye-straw.

~Varieties.~ These are almost as numerous as in the cabbage family. I find notes on some thirty-five varieties, tested from year to year, in my experimental grounds. Most of them prove themselves to be but a lottery, in this country of dry seasons, though in the moister climate of the European localities, where they are at home, they are a success.

The Half-Early Paris, or Demi-Dur, was for years the standard variety raised in this country, and from this, by selection, favorite local varieties were obtained; but, of late years, this has been, to a large degree, superseded by several excellent sorts, of which the Extra-Early Dwarf Erfurt was, doubtless the parent. Principal among these varieties are the Snowball, the Sea-Foam, Vick's Ideal, and Berlin Dwarf. All of these are early sorts and excellent strains. After testing them side by side, I find that the best strain of the Snowball is not excelled by either of them. Of the somewhat later ripening sorts, a variety which originated in this country, called the "Long Island Beauty," gives me great satisfaction, in its reliability for heading, and in the large size of its heads; this, with the Algerian, as a larger late sort, will give us a first-class series.

Cauliflower seed is not raised, as yet, to any large extent in this country, though some successful efforts have recently been made in this direction. I have found that there is a remarkable difference between varieties in the quantity of seed they will yield. From one variety I have raised as high as sixty pounds of seed from a given number of plants, while from two others, equally early, having the same number of plants in each instance, and raised in the same location (an island in the ocean), with precisely the same treatment in every way, I got, in each case, less than a tablespoonful of seed, though the heads of some of them grew to the enormous size of sixteen inches in diameter.

A fine cauliflower is the pet achievement of the market gardener. The great aim is not to produce size only, "but the fine, white, creamy color, compactness, and what is technically called curdy appearance, from its resemblance to the curd of milk in its preparation for cheese. When the flower begins to open, or when it is of a warty or frost-like appearance, it is less esteemed. It should not be cut in summer above a day before it is used." The cauliflower is served with milk and butter, or it may become a component of soups, or be used as a pickle.

The ~Broccoli~ are closely allied to the cauliflower, the white varieties bearing so close a resemblance that one of them, the Walcheren, is by some classed indiscriminately with each. The chief distinction between the two is in hardiness, the broccoli being much the hardier.

Of Broccoli over forty varieties are named in foreign catalogues, of which WALCHEREN is one of the very best. KNIGHT'S PROTECTING is an exceedingly hardy dwarf sort. As a rule, the white varieties are preferred to the purple kinds. Plant and treat as cauliflower.

Of ~Brussels-Sprouts~ (or bud-bearing cabbage) there are but two varieties, the dwarf and the tall; the tall kind produces more buds, while the dwarf is the hardier. The "sprouts" form on the stalks, and are miniature heads of cabbage from the size of a pea to that of a pigeon's egg. They are raised to but a limited extent in this country, but in Europe they are grown on a large scale. The sprouts may be cooked and served like cabbage, though oftentimes they are treated more as a delicacy and served with butter or some rich sauce. The FEATHER STEM SAVOY and DALMENY SPROUTS are considered as hybrids, the one between the brussels-sprouts and Savoy, the other between it and Drumhead Savoy. The soil for brussels-sprouts should not be so rich as for cabbage, as the object is to grow them small and solid. Give the same distance apart as for early cabbage, and the same manner of cultivation. Break off the leaves at the sides a few at a time when the sprouts begin to form and when they are ready to use cut them off with a sharp knife.

~Kale.~ Sea-kale, or sea-cabbage, is a native of the sea coast of England, growing in the sand and pebbles of the sea-shore. It is a perennial, perfectly hardy, withstanding the coldest winters of New England. The blossoms, though bearing a general resemblance to those of other members of the cabbage family, are yet quite unique in appearance, and I think worthy of a place in the flower garden. It is propagated both by seed and by cuttings of the roots, having the rows three feet apart, and the plants three feet apart in the rows. It is difficult to get the seeds to vegetate. Plant seed in April and May. The ground should be richly manured, and deeply and thoroughly worked. It is blanched before using. In cooking it it requires to be very thoroughly boiled, after which it is served up in melted butter and toasted bread. The sea-kale is highly prized in England; but thus far its cultivation in this country has been very limited.

The ~Borecole~, or common kale, is of the cabbage family, but is characterized by not heading like the cabbage or producing eatable flowers like the cauliflower and broccoli. The varieties are very numerous, some of them growing very large and coarse, suitable only as food for stock; others are exceedingly finely curled, and excellent for table use; while others in their color and structure are highly ornamental. They are annual, biennial, and perennial. They do not require so strong a soil or such high manuring as other varieties of the cabbage family.

The varieties are almost endless; some of the best in cultivation for table use are the DWARF SCOTCH, DWARF GREEN CURLED or GERMAN GREENS, TALL GREEN CURLED, PURPLE BORECOLE, and the variegated kales. The crown of the plant is used as greens, or as an ingredient in soups. The kales are very hardy, and the dwarf varieties, with but little protection, can be kept in the North well into the winter in the open ground. Plant and cultivate like Savoy cabbage.

The variegated sorts, with their fine curled leaves of a rich purple, green, red, white, or yellow color, are very pleasing in their effects, and form a striking and attractive feature when planted in clumps in the flower garden, particularly is this so because their extreme hardiness leaves them in full vigor after the cold has destroyed all other plants--some of the richest colors are developed along the veins of the uppermost leaves after the plant has nearly finished its growth for the season. The JERSEY COW KALE grows to from three to six feet in height and yields a great body of green food for stock; have the rows about three feet apart, and the plants two to three feet distant in the rows. In several instances my customers have written me that this kale raised for stock feed has given them great satisfaction.

The THOUSAND-HEADED KALE is a tall variety sending out numerous side shoots, whence the name.

* * * * *

www.ingramcontent.com/pod-product-compliance
Lightning Source LLC
Chambersburg PA
CBHW070332190526
45169CB00005B/1865